T0310228

Aircraft Performance

Aircraft Performance

Maido Saarlas
Professor
Department of Aerospace Engineering
U.S. Naval Academy

John Wiley & Sons, Inc.

This book is printed on acid-free paper. ♾

Published by John Wiley & Sons, Inc., Hoboken, New Jersey
Published simultaneously in Canada

For general information about our other products and services, please contact our Customer Care Department within the United States at (800) 762-2974, outside the United States at (317) 572-3993 or fax (317) 572-4002.

Wiley also publishes its books in a variety of electronic formats. Some content that appears in print may not be available in electronic books. For more information about Wiley products, visit our Web site at www.wiley.com.

Library of Congress Cataloging-in-Publication Data:

Saarlas, Maido.
Aircraft performance/by Maido Saarlas.
p. cm.
Includes bibliographical references and index.
ISBN-10: 0-470-04416-0 (cloth)
ISBN-13: 978-0-470-04416-2 (cloth)
1. Airplanes—Performance. I. Title.

TL671.4.S228 2006
629.132—dc22

2006043974

Printed in the United States of America

10 9 8 7 6 5 4 3 2

[To]
the University of Illinois, where it all started
[and]
my family who nurtured and supported me

Preface

This is intended to be a textbook mainly for aeronautical engineering students. Since college-level physics and integral calculus are the essential required background, the material in this book should also be accessible to other engineering students or as reference material to practitioners in the aerospace industry.

The approach is to minimize aerodynamics and propulsion-driven methodologies that often lead to extensive (but necessary) approximations and loss of overview of the problem. Thus, aerodynamics and propulsion are considered as known inputs to the mechanics equations, which then permit exposing the salient features of the performance problem at hand without getting lost in some very detailed issues. A minimal amount of aerodynamics and propulsion information, provided in the appendices, should provide (both a novice and an experienced hand) a relatively safe path for estimating essential performance features of an aircraft. The methods described in Chapters 2 through 7—level flight, climb, range, take-off, and maneuvering—are applicable for both nominal assessment as well as for extensive analysis needed for mission evaluation and design. The difference lies in aerodynamics and propulsion details (e.g., critical Mach number, high lift devices, etc.) and assumptions and approximations made to obtain practical results. Approximations become a roadmap that defines validity of the results and overlaps performance and design issues. Design, with roots in performance requirements, means additional approximations and assumptions pursuant to requirements for a design; a small change in requirements (e.g., from a 4g maneuver capability to 6g) may result in an entirely different aircraft design. Although this book is not about design, constraints for the performance-oriented design selection process is one of the topics in Chapter 8.

The central idea of this book is to maintain focus on basic aircraft performance. The central theme is the energy method.

The seeds of this book can be found in early energy method developments in the 1950s. Historically, the earliest works are by Lush, Kelly, and Rutowski (see bibliography), who developed and articulated a more fundamental approach to aircraft performance evaluation. Their work showed that use of the concept of total energy (kinetic + potential energies) and the rate of change of total energy as new independent variables (rather that the usual altitude or speed) enhanced both the understanding and quality of aircraft performance evaluation.

This immediately spawned further work on optimization of performance and flight paths, which provided results now regularly used in aircraft operations. Unfortunately, little of this approach could be found in the textbooks of 1960s and 1970s, as the level of mathematics involved implied graduate level of involvement or the calculations were viewed as excessively computer driven and cumbersome.

Thus, this set of notes was started with two very general goals in mind:

1. To include new and evolving approaches for performance analysis, starting with the basic energy concepts
2. To minimize the aerodynamic and propulsion-driven methodologies (i.e., to treat aerodynamics and propulsion only as necessary inputs for the mechanics equations)

Obviously, the second goal needs to be modified when high-speed aerodynamics and supersonic inlets require more detailed and specialized treatment. The basic methods still remain valid.

Energy method provides a somewhat unified approach, and it works well over a practical subsonic-supersonic speed range. Performance is approached from the point of energy production and balance culminating in one fundamental performance equation. This provides a point of departure for handling practical point performance problems. But the energy method can also handle the quasi-dynamic path performance problems arising from minimum-time and minimum-fuel consumption requirements. Rutowski's work allows now evaluation of those basic optimization problems with acceptable accuracy, as compared to complete and precise numerical optimization process.

Part of this success is due to evolution of (hand) calculators. The available functions and solver processes can make short work now of the arctan function in range problems (see Chapter 5) and fourth-degree polynomials regarding maximum velocity (see Chapter 3). Programmable features reduce also the optimum path problems to a few elementary steps. However, this does not necessarily obviate the classical graphical approach described in Chapter 3. It arose from necessity to circumvent the cumbersome trial-and-error solution of the fourth-degree polynomial for maximum velocity. It still provides a quick overview of the entire performance potential, and is also easily accessible via new graphing routines.

These notes have been used in late 1950s at UCLA, in the 1960s at the University of Cincinnati, and since 1970 at the U.S. Naval Academy. They have been continuously modified, set aside for every new text that appeared, and picked up again to pursue the two goals previously outlined. They have been updated, streamlined, and made more relevant by the large number of military pilots and faculty who have passed through this department. They are too numerous to mention individually, and lest a good friend will be omitted, they all should know that the author owes them deep gratitude.

There is more material in these notes than can be covered comfortably in one semester. It has been found that the sections marked with an asterisk (*) can be omitted without loss of continuity and without sacrificing understanding of the topic at hand. Each chapter contains numerical examples drawn from practice and numerous problems; some with answers. Appendices A through E contain altitude tables and data on aircraft and propulsion.

Contents

1

The General Performance Problem

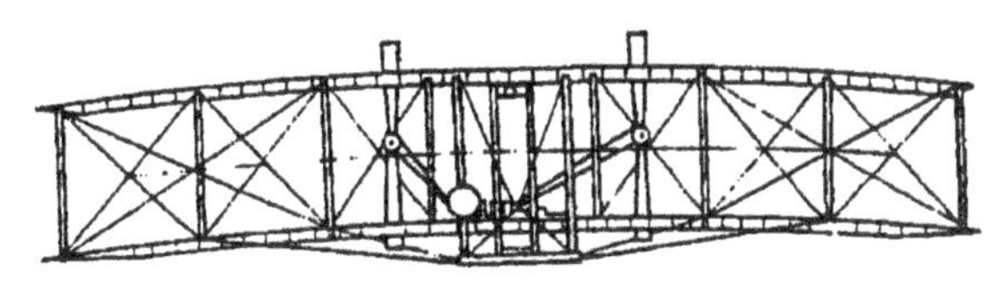

Wright Flyer

1.1 INTRODUCTION

An engineer-designer who decides on the configuration, size, arrangement, and the choice of power plant for an air vehicle must make the decision on the basis of the performance that will be expected of the finished product. For this reason, one must be familiar with the basic performance characteristics and with the relationship of design factors that can influence these characteristics.

In addition to the direct design needs, an accurate knowledge of aircraft performance is necessary for the operator of the aircraft. The airlines need this information to determine how the aircraft can be operated most efficiently and economically. Similarly, the armed services need to know what a proposed or given aircraft can do, and how it must be flown in order to gain the best possible advantage or provide most effective support.

Thus, an engineer needs a good understanding of aircraft performance problems for use both in preconstruction design and in analysis or evaluation of the finished aircraft. Similarly, a knowledge of possible performance characteristics and limitations of various classes of aircraft is needed to establish sound strategy in both commercial and military operations.

This book deals with the methods by which the performance of an aircraft can be determined from its aerodynamic and powerplant characteristics. For present purposes, it will be assumed that the aerodynamic and propulsive data are known and given, and for clarity the main topic will be performance. This implies that performance does not affect aerodynamics or propulsion. Strictly speaking, this is not true because different performance regimes (i.e., speed, range, altitude) may require different wing sweep, aspect ratio, or power plant—which, in turn, establishes different performance characteristics. However, such a design iteration procedure is outside the scope of these notes. Moreover, it is not intended to treat aerodynamics, propulsion, and performance under the same cover, as each of these are independently worthy topics requiring a substantial amount of space for adequate coverage. In order to facilitate working example problems, and for otherwise handy reference, special appendices cover some aerodynamic and propulsive equations and data.

1.2 PERFORMANCE CHARACTERISTICS

This book does not attempt to solve the complete dynamic flight mechanics problem where the translational and rotational motion of the vehicle is solved along a trajectory. Such a problem formulation requires a set of simultaneous differential equations describing the forces, moments, and the attitude of an aircraft as functions of position and time. In general, only numerical solutions are then possible.

To preserve clarity and simplicity, and to provide analytical simplifications leading to a good overview of the physical problem involved, a quasi-steady approach will be used. The flight path will be confined to either horizontal or vertical plane at a time over a "flat earth." As a result, cross-coupling and inertial terms can be neglected in the dynamic equations of motion. Since the attitude of the vehicle is of secondary interest in performance calculations, it will be assumed that the forces can be considered independently of the moments. This then permits ignoring the moments and the attitude of the vehicle, and the flight dynamics problem is reduced to considering the translational motion of a *point mass* with a variable mass due to propellant consumption. The performance problem then means studying various aspects of translational motion, such as how far the vehicle will fly, how fast, and so on.

From a mathematical point of view, one finds two basic problems in aircraft performance: point performance and path performance (integral performance). Point performance problems describe the particular local performance characteristics at a given point on the path independent of the rest of the flight path. The path performance deals with the study of the flight path as a whole, and involves integration between given initial and final conditions along the flight path (i.e., range, time to climb).

In general, the point performance problem represents a set of algebraic equations obtained from the dynamic equations by neglecting the time-dependent terms. Thus, one can find the local extrema by elementary techniques of the differential calculus. The path performance problems usually require integration along the flight path and, without resorting to numerical methods, can be solved only with most restrictive assumptions concerning either altitude or velocity, or both. The determination of optimum flight conditions requires techniques of the calculus of variations.

Although both point and path performance problems will be discussed in the following chapters, another more functionally oriented division of flight performance problems transcends the two approaches. From the point of view of the dual purpose of performance calculations outlined in Section 1.1, it is natural to divide the performance characteristics of an aircraft into two groups as follows: absolute performance characteristics and functional performance characteristics.

1.2.1 Absolute Performance Characteristics

The first group, called the *absolute performance characteristics,* are of the primary interest to the designer. They are called absolute because they are simply numbers representing the capabilities of the aircraft. The most important are:

- Maximum speed
- Stalling speed
- Best climbing speed
- Best glide speed

- Rate of climb
- Ceiling(s)
- Maximum range and speed for maximum range
- Maximum endurance and speed for maximum endurance
- Take-off distance
- Landing distance

1.2.2 Functional Performance Characteristics

The second group of performance characteristics will be called *functional performance characteristics,* as they are more important from the standpoint of the efficient operation of the aircraft. These characteristics are not expressed simply as numbers, but rather as functions or curves (i.e., such as the variation of airspeed with altitude necessary to accomplish a certain purpose). Typically, the characteristic functions constitute the answer to the following questions:

1. What is the program of speed and altitude that must be followed in order to go from a given altitude h_1 to another altitude h_2 in minimum time?
2. What is the program of airspeed and altitude to follow in order to go from one flight condition (i.e., speed and altitude), to another in minimum time?
3. What is the program of altitude and speed such that the aircraft can change from one flight condition (i.e., speed and altitude) to another with minimum expenditure of fuel?
4. What variation in flight conditions will permit the aircraft to cover the greatest distance over the ground?

It is easily seen that the so-called absolute characteristics in the first group are design-compromise and specifications oriented, as there is a direct connection between these and the variables describing the aircraft geometry, weight, and the power plant. The second group consists of questions of great importance to successful commercial or tactical use of the aircraft.

Last, and by no means least, it should be realized that the individual performance characteristics, obtained by whatever means, must be combined into a unified description of the desired or specified flight profile. The study of a flight profile and the requirements it places on the performance characteristics, and vice versa, is called mission analysis. Although the mission analysis often precedes the detailed performance calculations, as it is used as a tool for generating the general aircraft specifications, it also relies on the techniques used for performance analysis. A full treatment of the mission analysis is beyond the intent and scope of these notes, however some discussion can be found in Chapter 8, section 8.1.

1.3 THE APPROACH

Typically, a sound engineering approach to problem solving is to draw a free body diagram, consider all the forces and moments acting on it, and then apply all appropriate conservation laws. In a true dynamic environment, additional information must be provided concerning kinematics, aircraft attitude, and a reference system that helps to locate the vehicle. As already pointed out, such an approach produces complete and accurate information about the behavior of the vehicle in the form of numerical solution. For many engineering purposes, analytical solutions—even if approximate—are desired, as they provide quick overview, trends, and adequate numerical information.

Such an approach is followed here. Emphasis is on accessible and useful solutions with reasonable and sensible accuracy. The general three-dimensional problem is divided into two separate problems located in horizontal and vertical planes. Equations of motion, based on the free body diagram, are obtained for each separate plane. This standard approach is provided for flexibility, familiarity, and ease of solution of a number of point performance problems. It is also helpful for understanding and supporting the energy method used for most of topics at hand.

The main theme approach is to apply the concept of energy production and balance to aircraft performance evaluation. A single energy equation, fundamental performance equation (FPE), will be developed and related to existing equations of motion. This serves to clarify and amplify some standard point-performance methods. It also leads to quasi-steady maneuvering analysis and is a central function in solutions of some path performance problems.

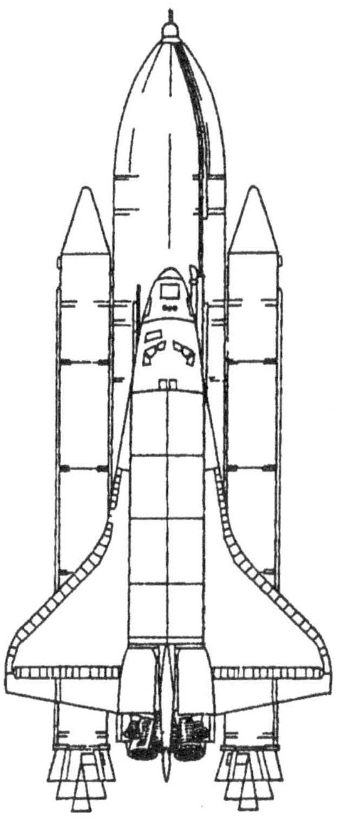

Shuttle

2

Equations of Motion

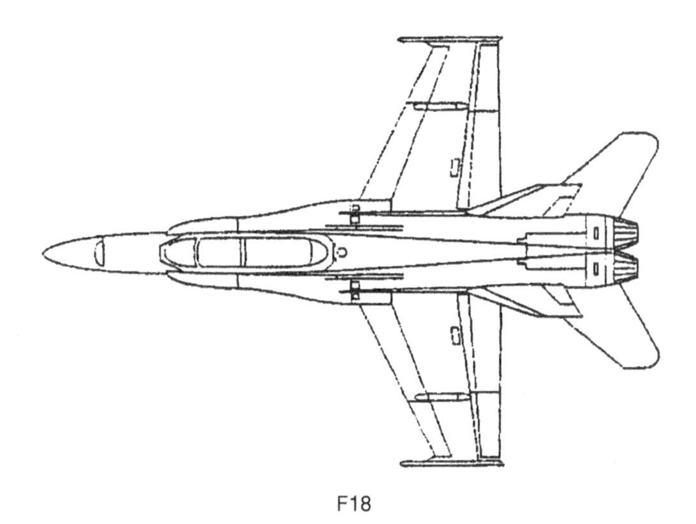

F18

2.1 GENERAL INFORMATION

Before launching into performance analysis via the energy equation it may be beneficial to review the standard dynamics approach that consists of a free body diagram of a point mass aircraft. They both should give the same answers. When and why they diverge is the answer sought in this book. Since the predominant portion of flight time is spent either in almost rectilinear flight at constant altitude or in change from one altitude to another, it is appropriate to consider first the flight that is confined entirely to a vertical plane defined by altitude h and horizontal distance x. Combined with the mathematical simplifications the physical picture can then be maintained at such a level that practical and significant results are easily obtainable. Flight in the horizontal plane—maneuvering flight and high-g turns—will be considered later.

The system of equations describing the motion is obtained from Newton's second law

$$\vec{F} = m\frac{d\vec{V}}{dt} \tag{2.1}$$

and the forces acting on the flight vehicle, as presented in Figure 2.1. The total force $\vec{F}$ consists of the aerodynamic force $\vec{A}$, thrust $\vec{T}$, and the gravitational term $m\vec{g}$. $\vec{A}$, in turn, is given in terms of lift L and drag D. By convention, lift acts perpendicular to the velocity vector $\vec{V}$ and the drag is along and in the direction opposite to $\vec{V}$. Thus, Eq. 2.1 becomes:

$$\vec{A} + \vec{T} + m\vec{g} = m\frac{d\vec{V}}{dt} = m\dot{\vec{V}} \tag{2.2}$$

Since the flight path is in the vertical plane, Eq. 2.2 can be written in the component form normal, $\vec{n}$, and tangent, $\vec{t}$, to the flight path. With the definition of the aerodynamic force

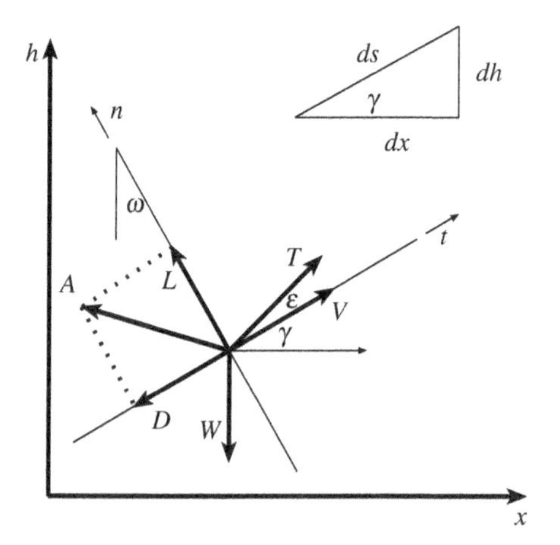

Figure 2.1 Flight Path in a Vertical Plane

$$\vec{A} = L\vec{n} + D\vec{t} \tag{2.3}$$

and the acceleration normal to the flight path

$$a = V\frac{d\gamma}{dt} = V\dot{\gamma} \tag{2.4}$$

one obtains the resolved equations in $\vec{t}$ and $\vec{n}$ directions, respectively:

$$-D + T\cos\epsilon - mg\sin\gamma = m\dot{V} \tag{2.5}$$

$$L + T\sin\epsilon - mg\cos\gamma = mV\dot{\gamma} = mV\omega \tag{2.6}$$

where γ is the angle of the flight path with the horizontal plane. The dot superscript signifies differentation with respect to time. ω is the rotation rate of the local radius vector of the path and is given by $\dot{\gamma}$, the rate of change of inclination of the flight path to the horizontal. The velocity $\vec{V}$ is tangent to the flight path.

In addition, the following kinematic relationships hold along the flight path:

$$\frac{dx}{dt} = \frac{ds}{dt}\cos\gamma = V\cos\gamma \tag{2.7}$$

$$\frac{dh}{dt} = \frac{ds}{dt}\sin\gamma = V\sin\gamma \tag{2.8}$$

Eqs. 2.5 through 2.8 complete the set of equations of motion in a vertical plane.

In general, the thrust T acts at a very small angle along the flight path, and is a function of velocity and altitude. The lift L and drag D are functions of altitude, velocity, and angle of attack α. Since the lift is directly a function of angle of attack, and since the latter does not appear explicitly in the formulation of Eqs. 2.5 to 2.8, it is convenient to replace α by the lift L. The equations of motion can then be written in the form with ($W = mg$):

$$\dot{h} = V\sin\gamma \tag{2.9}$$

$$\dot{x} = V\cos\gamma \tag{2.10}$$

$$\frac{\dot{V}}{g} = \frac{[T(h, V) \cos \epsilon - D(h, V, L)]}{W} - \sin \gamma \tag{2.11}$$

$$\frac{\dot{\gamma}}{g} = \frac{[T(h, V) \sin \epsilon + L(h, V)]}{VW} - \frac{\cos \gamma}{V} \tag{2.12}$$

$$\dot{m} \equiv -\frac{dW}{dt} \tag{2.13}$$

It is easily recognized that Eq. 2.9 represents the velocity in the vertical direction—the rate of climb—and that Eq. 2.10 leads, by integration, to the range R ≡ x. Eq. 2.13, an identity, states that the change of mass is due to the change of the weight of the aircraft.

A simple inspection reveals that there are a total of six possible variables (m, h, V, γ, L, x) and only four equations. Thus, additional simplifications must be made or the equations will lead to a two-parameter family of solutions. From the practical point of view, this simultaneous set of differential equations is seldom solved for a parametric solution. Sufficient approximations are made that then lead to an algebraic set of equations for various rates (dh/dt, etc.), and only the range equation remains to be integrated. Such approximations, although extremely practical and useful, can lead to serious errors, especially in high-speed flight, and thus should be introduced with some care.

The solution process falls into two general categories. For the direct problem, the aerodynamic and thrust terms are known and the solutions are obtained for the kinematic variables (range, rate of climb, maximum velocity, etc.). If the kinematic results are known, then the required aerodynamic or thrust data are sought to satisfy the specified conditions. This is the inverse of the design problem, and the solutions are obtained—often in parametric form.

This text will eventually introduce and follow most of the usual assumptions and methods, but the approach used here will be somewhat different from those used in the standard textbooks. The process can be broken down into three overall steps:

1. The basic performance will be formulated in terms of the aircraft total energy variation.
2. The standard steady-state problem will be deduced as a number of special cases permitting a clear overview of all pertinent assumptions used, the range of applicability, and need for energy consideration as the flight speed increases.

3. The energy formulation will lead to a number of optimal performance problems that can provide engineering solutions without recourse to extensive computational equipment.

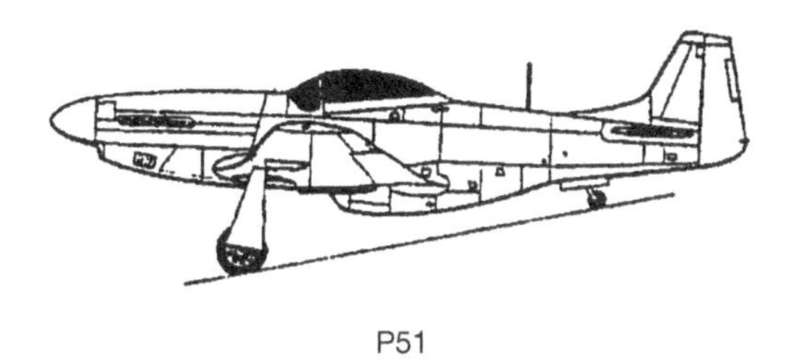

P51

2.2 THE ENERGY APPROACH

In formulation of a number of flight performance problems, it is convenient, and perhaps necessary, to use the energy balance rather than the force equations. As the resulting equation represents a restatement of the force equations, it is illuminating to investigate the energy equation and the conditions that must be satisfied for its use independent of the rest of the force equations.

The energy balance is essentially a statement about the source of the energy and how the energy is spent. The source is the fuel used, which is characterized by its energy content. The fuel/chemical energy is converted into useful work by the aircraft powerplant at an overall efficiency η_o. The energy content of the fuel H_f and fuel (mass) flow rate $(dW_f/g)/dt$ give the power available to the engine:

$$P_a = \eta_o \frac{H_f}{g} \frac{dW_f}{dt} \tag{2.14}$$

This power is spent to overcome the drag and to provide energy for the aircraft performance. Thus, the following energy balance must be satisfied:

Rate of change of aircraft energy	=	Power available from fuel	−	Power dissipated by drag	+	Rate of energy change due to weight change of aircraft

The total energy of the aircraft E is given by the sum of its potential and kinetic energies as

$$E = W\left(h + \frac{V^2}{2g}\right) = We \tag{2.15}$$

where e is the specific energy, V is the flight speed, and h is the altitude:

$$e = h + \frac{V^2}{2g} \tag{2.16}$$

Thus, the rate of change of aircraft specific energy may be written as

$$\frac{de}{dt} = \frac{dh}{dt} + \frac{V}{g}\frac{dV}{dt} \tag{2.17}$$

The energy balance may then be expressed as

$$\frac{dE}{dt} = P_a - DV + \frac{edW}{dt} \tag{2.18}$$

Differentiating now Eq. 2.15

$$\frac{dE}{dt} = \frac{W\,de}{dt} + \frac{edW}{dt} \tag{2.19}$$

and substituting into Eq. 2.18, one obtains the aircraft basic energy equation, which is used throughout this book:

$$\frac{P_a - DV}{W} = \frac{dh}{dt} + \frac{V}{g}\frac{dV}{dt} = \frac{de}{dt} \equiv P_s \tag{2.20}$$

Eq. 2.20 shows that the rate of change of specific energy, e, is given by excess (or lack of) power per pound of aircraft weight. More important, the left-hand side may be viewed as the *energy generation term,* where power available is balanced against power dissipated by drag. The right-hand side may be considered as *energy distribution* between altitude and velocity changes. Either side may be zero. The left-hand side vanishes if all the power available is used to overcome drag. de/dt may be zero for a variety of combination of altitude and velocity variation. The commonly accepted symbol P_s, which may represent either side of the energy equation (see Chapters 7 and 8), is called specific excess power.

Eq. 2.20 is a powerful tool in the study of efficient and comparative aircraft maneuvering. Knowing the specific excess power determines the aircraft climb and acceleration capabilities. Conversely, knowing (or specifying) rate of climb and/or acceleration leads to knowledge (requirements) for aircraft power and drag levels (i.e., the excess power).

Dimensionally, Eq. 2.20 represents a velocity. This is easily verified if one recalls the definitions of potential and kinetic energies. Potential energy is the work done by a force to raise the vehicle to height h, or

$$P.E. = Fh = mgh = Wh\ [lb_f\text{-}ft],\ [N\text{-}m],\ [J] \tag{2.21}$$

Kinetic energy is the work done by the resultant force to bring a vehicle from rest to motion with a speed V, or

$$K.E. = Fx = (ma)\left(\frac{1}{2}at^2\right) = \frac{1}{2}ma^2t^2 = \frac{1}{2}mV^2 \tag{2.22}$$

Thus, the units of energy E: $[lb\text{-}ft]$ or $[N\text{-}m]$ and for specific energy, e: [ft], [m] where $e = E/W$. Then, it follows that

$$\frac{de}{dt} = \frac{P - DV}{W}\left[\frac{\text{ft}}{\text{sec}}\right] \text{ or } \left[\frac{\text{m}}{\text{sec}}\right] \tag{2.23}$$

As examples of specific applications of Eq. 2.20, consider the unaccelerated level flight where h = constant, V = constant, and $de/dt = 0$. Then one obtains:

$$P_a = DV = P_r \tag{2.24}$$

where P_r is the power required to sustain equilibrium level flight. Eq. 2.24 is the basic static performance statement.

As a second example, if one assumes flight at constant velocity, then it follows from Eq. 2.17 that $de/dt = dh/dt$, and Eq. 2.20 reduces to the quasi-steady rate of climb equation:

$$\frac{dh}{dt} = \frac{P_a - DV}{W} \tag{2.25}$$

To place the energy equation, Eq. 2.20, into a different perspective, it can be derived also from the force equations, Eq. 2.11 and Eq. 2.9. To this end, Eq. 2.11 will be written

$$\frac{\dot{V}}{g} + \sin\gamma = \frac{T - D}{W} \tag{2.26}$$

And using Eq. 2.9, we get:

$$\dot{h} + \frac{\dot{V}V}{g} = \frac{(T - D)V}{W} \tag{2.27}$$

The left side can immediately be recognized as de/dt, and since $TV = P_a$, one recovers the energy equation, Eq. 2.20.

At this point it seems that the energy equation, after being first derived independently of the other path equations, could be used as an independent equation. However, it is coupled to the other path equations through lift and the flight path angle γ (see Eqs. 2.9 to 2.12). Since these equations represent a set of simultaneous nonlinear differential equations, certain assumptions must be made before individual equations can be treated as such.

There are three sets of assumptions, called *energy assumptions,* that will effectively decouple these simultaneous equations, Eqs. 2.9 to 2.12:

1. Assuming that the flight path angle is constant, or that $\dot{\gamma}$ is very small, with $\epsilon \simeq 0$, the lift can be given independent of the path angle and thrust, Eq. 2.12. Thus, Eq. 2.11 becomes quasi-independent.
2. Drag is only a function of altitude and velocity. Then $\sin\gamma$ can be eliminated as above, and one obtains the energy equation from Eq. 2.11.
3. Neglecting $\dot{V}$, or both inertia terms: $\dot{V}$ and $\dot{\gamma}$, decouples the equations.

The merits and disadvantages of these assumptions become evident in those specific sections where individual performance problems are treated. It is sufficient to point out at this time that these assumptions, although somewhat restrictive, are necessary in order to establish performance solutions for engineering expediency and accuracy. Moreover, a basic aspect of performance analysis is establishing (at least to first order) some optimum performance conditions. Without the use of some of the assumptions and the attendant simplification of the equations, the optimizing process becomes extremely cumbersome and amenable only to computer methods.

PROBLEMS

2.1 What is the potential energy of a 40,000 lb aircraft at 40,000 ft above sea level?

2.2 If the aircraft in Problem 2.1 is flying at 1,000 ft/sec, what is its kinetic energy? Its total energy?

2.3 If the aircraft in Problem 2.1 descends to 20,000 ft and continues flying at 1,000 ft/sec, what is its energy now? What happened to the energy difference?

2.4 The average heat content of JP-4 fuel or aviation gasoline is 19,000 BTU/lb. Calculate the power available if the efficiency of the engine is 20 percent and the fuel flow rate is 750 lb/hr. Assume a typical propeller efficiency to be 0.85.
Ans: $P_a = 952$ HP.

3

The Basics

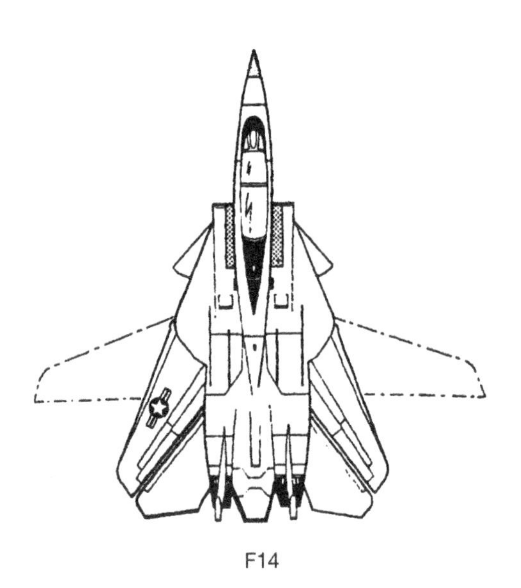

F14

3.1 FUNDAMENTAL PERFORMANCE EQUATION

The general types of flight performance problems discussed in Chapter 1 were divided into two basic groups: absolute and functional performance characteristics. In addition, analytical development of the equation of motion and the nature of the motion indicates that most problems can be divided into quasi-steady (static) or unsteady (dynamic) categories. Some, like range and endurance, can be treated by either approach, depending on the assumptions used. It turns out that most absolute performance characteristics can be treated mathematically as steady-state point-performance problems leading to simple algebraic expressions. Local (path independent) optimum conditions are determined by simple differential calculus. The dynamic problems, which arise mainly from variation of velocity and altitude along the flight path, deal with functional performance characteristics and the transient behavior of the aircraft. Their solution requires the consideration of motion along the entire flight path, which leads to use of extensive numerical methods. Some practical simplifications are found in Chapter 8.

The purpose of this chapter is to establish suitably simple engineering expressions, while determining which data are required for evaluating the thrust T, drag D and power P. To that end, the main analysis vehicle is the basic energy equation Eq. 2.20, which is used to focus on those performance characteristics that establish the general bounds for powered flight vehicle performance in steady nonmaneuvering flight: stalling speed, maximum and minimum speeds, and ceiling. These are often called the basic performance problems, and they define a theoretical region on an altitude-velocity plot where $T \geq D$ and where steady-state flight is possible. This plot is called aircraft *flight envelope* and will be established in subsequent sections.

For unpowered vehicles (gliders, space shuttles), glide performance is important and will be discussed as the last section in this chapter. Stalling speed and glide problems, in their essential features, are independent of the power plant and are essentially functions of aerodynamics. However, power does influence both stall and glide performance and may need to be considered in full aircraft performance evaluation and design. As will be seen shortly, maximum and minimum speeds and ceiling are obtained at full thrust (power) setting and thus are functions of the powerplant. Consequently, their solutions proceed according to whether the engine characteristics are thrust (jet engines) or power (propellers) oriented.

Before commencing discussion of the performance problems, the energy equation will be recast in a different form suitable for direct performance calculations. The resulting equation is the *Fundamental Performance Equation* (*FPE*), which will be used for most of the development in subsequent chapters.

This FPE is obtained by rewriting the energy equation:

$$\frac{P_{\mathrm{a}} - DV}{W} = \frac{dh}{dt} + \frac{V}{g}\frac{dV}{dt} = \frac{de}{dt} \equiv P_{\mathrm{s}} \tag{3.1}$$

Solving for the rate of climb yields

$$\left[\frac{dh}{dt} = \frac{P_{\mathrm{a}} - DV}{W}\right] - \frac{V}{g}\frac{dV}{dt} \tag{3.2}$$

where P_{a} is the power available from the engine. Since P_{s} is just a convenient shorthand symbol for either side, it plays no role in this development. Its usefulness is demonstrated in Chapters 7 and 8.

The terms inside the square brackets constitute the *classical fundamental performance equation* where the assumption has been used that the climb or descent velocity is constant, or, $dV/dt = 0$ (see also Eq. 2.25). It is classical in the sense that when used in early low-speed flight analysis, ignoring the acceleration term did not introduce significant errors and the bracketed term was easily derivable from simple thrust-drag equilibrium. With increasing engine sizes and available steep climb angles, the acceleration term became significant, as it represented rapid energy change of the aircraft with attendant changes in velocity and/or altitude.

Considering Eq. 3.2 as representing excess (or lack) of power per pound of weight, several interpretations are possible (see Eqs. 2.24 to 2.25), as a power-drag unbalance will lead to changes in specific energy. The classical performance equation has its limitations, since only part of the energy balance is included. One way to improve accuracy for high speed or steep climb problems has been to establish a correction to the classical equation by rewriting Eq. 3.2 with

$$\frac{dV}{dt} = \frac{dV}{dh}\frac{dh}{dt} \tag{3.3}$$

as

$$\frac{dh}{dt} = \frac{P_{\mathrm{a}} - DV}{W}\frac{1}{\left[1 + \dfrac{V}{g}\dfrac{dV}{dh}\right]}. \tag{3.4}$$

The term in square brackets is then evaluated as a kinetic energy correction factor to the fundamental equation. For low speeds, the correction factor is small, as expected, but can exceed 2 for high speeds. As a result, the rate of climb dh/dt predicted by the classical formula may be actually halved. Among many other parameters, the correction depends on the climb technique and the powerplant characteristics. This is discussed in some detail later in the chapter.

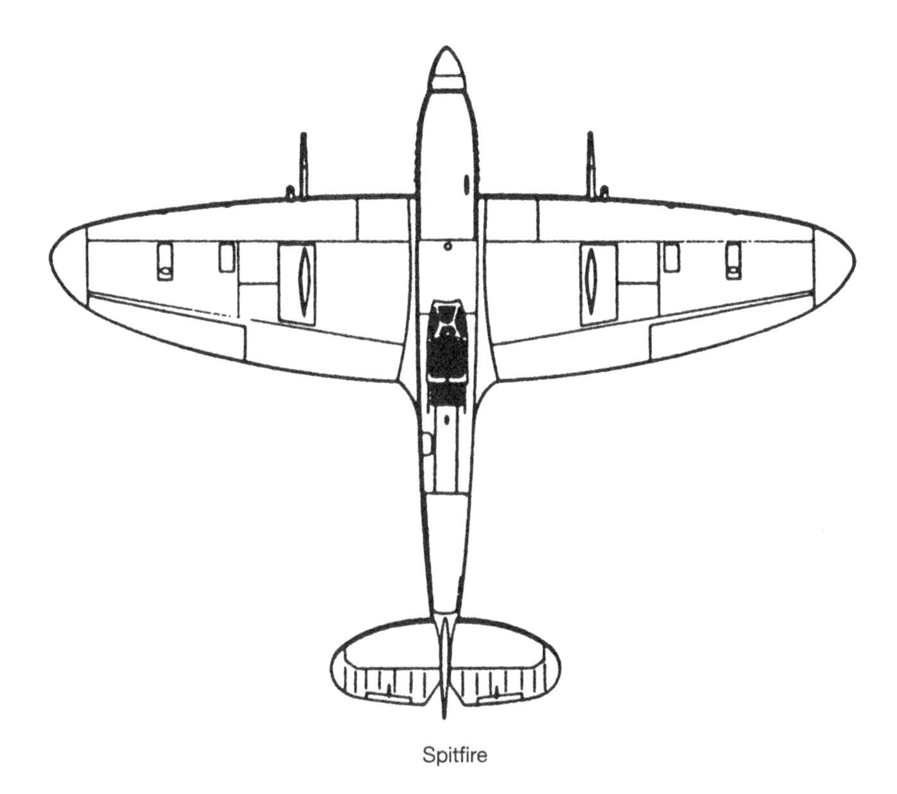

Spitfire

3.2 STALLING SPEED

The requirements for equilibrium flight are that the aircraft be flown at least at the speed that generates sufficient lift to counteract the aircraft weight and that the thrust be equal to drag. This is called the *stalling speed.* It occurs at the stall angle, or the maximum angle of incidence, and it represents the low end of the steady-state flight spectrum of the aircraft. For steady, level flight, equlibrium equations become (from Fig. 2.1 with $\gamma = 0$):

$$T = D$$

$$L = W$$

where, for simplicity, the thrust angle ϵ has been set equal to zero.
Lift equation, written in the usual form, is:

$$L = \frac{1}{2}\rho V^2 C_L S = W \tag{3.5}$$

where

ρ—density $\left[\frac{\text{slugs}}{\text{ft}^3}\right], \left[\frac{\text{kg}}{\text{m}^3}\right]$
V—velocity $\left[\frac{\text{ft}}{\text{sec}}\right], \left[\frac{\text{m}}{\text{sec}}\right]$
C_L—lift coefficient, dimensionless
S—wing reference area (ft^2), (m^2)
L—lift force (lb), (N)

This equation is also seen to be the simplest case of the equations of motion in Chapter 2 and represents the force balance normal to the flight path (Eq. 2.12) in level, unaccelerated flight. The thrust is assumed to be parallel to the path with no normal components, and the lift is assumed to be independent of drag and determined only by attitude, which also fixes the maximum value of lift. Thus, the velocity for steady, level flight is obtained from Eq. 3.5:

$$V = \sqrt{\frac{2W}{\rho S C_L}} \equiv TAS \tag{3.6}$$

It represents the relationship between the flight velocity and the required lift coefficient. It is also called true airspeed TAS. Eq. 3.6 also establishes the well-known relationship

$$V^2 C_L = \frac{2W}{\rho S} = \text{const}$$

for a constant altitude (and fixed W) flight.

The force balance along the flight path, Eq. 2.11, yields

$$T = D = \frac{W}{(L/D)} = \frac{W}{(C_L/C_D)} \tag{3.7}$$

where C_D is the dimensionless drag coefficient defined by

$$D = \frac{1}{2}\rho V^2 C_D S = T_r \tag{3.8}$$

Eq. 3.7 is the thrust required for level unaccelerated flight. Eq. 3.5 shows that, for a given altitude, the lift depends only on the lift coefficient and the velocity. However, for a fixed weight and altitude, a flight attitude exists that gives a maximum value to the lift coefficient $C_{L_{\max}}$ (although the latter depends also on lift augmentation devices like flaps and slats), which, in turn, indicates a minimum flight velocity in order to satisfy the above equation. This minimum velocity is called the stall speed, and is given by

$$V_s = \sqrt{\frac{2W}{\rho S C_{L_{\max}}}} \tag{3.9}$$

Eq. 3.9 represents the minimum flight velocity at which steady sustained flight is possible. It depends on the altitude, maximum lift coefficient, and, to some extent, on power at high angle of attack that must be determined experimentally. See also Problem 3.13.

It is convenient now to introduce the equivalent airspeed V_E or EAS, which is defined as

$$\frac{1}{2}\rho V^2 = \frac{1}{2}\rho_0 V_E^2 \tag{3.10}$$

where ρ_0 is the sea-level density. The equivalent airspeed establishes equivalence of the dynamic pressure at sea level and at altitude, which renders lift and drag characteristics the same at sea level and at altitude. Moreover, the standard air-speed indicator is theoretically calibrated to read equivalent airspeed. In practical performance calculations, use of V_E eliminates the altitude effect from many equations. Introducing V_E in Eq. 3.9 gives

$$V_{E_s} = \sqrt{\frac{2(W/S)}{\rho_0 C_{L_{max}}}} \tag{3.11}$$

Eq. 3.11 states that, for a given wing loading W/S, the equivalent stall speed is a function of the maximum lift coefficient (*attitude*), for all altitudes. The significance of this result is that, for a given aircraft, $C_{L_{max}}$ can be determined once and for all for any given combination of flaps and/or slats, and V_{E_s} can be calculated. Since, for a reasonably good instrument and installation, indicated airspeed (IAS) is almost equal to EAS, the aircraft will always stall at approximately the same IAS, which knowledge is of practical use to the pilot. Or, in other words, for a given aircraft, V_{E_s} is then a function of aircraft weight W only.

The indicated airspeed is the actual instrument indication in the aircraft at any given flight condition and altitude. A number of factors contribute to the difference between the figure shown by the IAS dial and the actual true airspeed (TAS): instrument error, installation error, position or instrument location error, compressibility at high speed, and deviation of local density from that of the sea level value ρ_0. The calibrated airspeed is

$$\begin{aligned} &\pm \text{ instrument error} \\ \text{CAS} = \text{IAS} &\pm \text{ installation error} \\ &\pm \text{ position error} \end{aligned}$$

and, in turn,

$$V_E \equiv \text{CAS} - \text{compressibility correction}$$

The standard airspeed indicator is calibrated to give a correct reading at standard sea-level conditions (14.7 psia, 59 F, 1 atm, 15 C). Since most flights take place in nonstandard conditions, corrections are made to IAS such that EAS represents that flight speed at standard sea level (std sl) with the same free stream dynamic pressure as the actual flight condition (Eq. 3.9). Finally, the true airspeed (TAS) is obtained when the EAS is corrected for density altitude (see Appendix A for atmospheric properties data):

$$V = TAS = \frac{EAS}{\sqrt{\sigma}} = \frac{V_E}{\sqrt{\sigma}} \tag{3.12}$$

where σ is the density ratio ρ/ρ_0.

Since the density depends on both pressure and temperature, and since CAS is corrected for pressure altitude, local sea-level pressure and temperature are needed to determine TAS from EAS (or CAS).

Theoretically, Eq. 3.11 also establishes the minimum velocity (at $C_{L\max}$) at which the aircraft can land. Practical flight safety rules, however, require that a landing speed above V_E be used. In general, the following requirements have been established:

$$V_L = 1.2\ V_s \text{ – visual approach}$$
$$V_L = 1.15\ V_s \text{ – carrier landing}$$
$$V_L = 1.3\ V_s \text{ – instrument approach}$$

These conditions then determine the required lift coefficients for respective landing conditions. The decreased lift coefficients and, therefore, the angle of attack, are reduced to provide required safety margins. In practice, this set of rules effectively has two purposes. First, it sets a minimum flight speed for safe landing. Second, it establishes the thrust requirements for safe landing (and take-off).

The following example explores the concepts of stall speed and $C_L - V$ relationship.

EXAMPLE 3.1

A Boeing 737 aircraft has the following characteristics:

$$W_{\max} = 111{,}000 \text{ lb}$$

$$W_{\max_{\text{land}}} = 103{,}000 \text{ lb}$$

$$S = 980 \text{ ft}^2$$

$$C_{L_{\text{land}}} = 2 \text{ at sea level}$$

$$V_{\max} = 850 \text{ ft/sec at } 40{,}000 \text{ ft}$$

Calculate:

a. V_s
b. $C_{L\max}$
c. The relationship between the lift coefficient and the velocity over the usable speed range at the landing weight at 40,000 feet, if the safe landing speed is taken as 1.2 V_s

a. Landing speed is calculated from

$$V_L = \sqrt{\frac{2W}{\rho_0 S C_L}} = \sqrt{\frac{2 \times 103{,}000}{0.002377 \times 980 \times 2}} = 210 \, \frac{\text{ft}}{\text{sec}}$$

Thus

$$V_s = \frac{210}{1.2} = 175 \, \frac{\text{ft}}{\text{sec}}$$

b. The lift coefficient at stall, $C_{L_{\text{max}}}$, can be calculated, because the weight and altitude are fixed.

$$C_{L_{\text{land}}} \times V_{\text{land}}^2 = C_{L_{\text{max}}} \times V_s^2 = \text{const}$$

$$C_{L_{\text{max}}} = 2 \, \frac{V_{\text{land}}^2}{V_s^2} = 2 \times 1.44 = 2.88$$

c. Since at sea level EAS = TAS, the stall speed at 40,000 feet can be obtained from

$$V_{E_s} = 175 \, \frac{\text{ft}}{\text{sec}}$$

$$V_{s\,40{,}000} = \frac{V_{E_s}}{\sqrt{\sigma}} = \frac{175}{\sqrt{0.247}} = 352 \, \frac{\text{ft}}{\text{sec}}$$

Thus, the usable velocity range at 40,000 ft is

$$352 \, \frac{\text{ft}}{\text{sec}} \le V \le 850 \, \frac{\text{ft}}{\text{sec}}$$

The relationship between the lift coefficient and the flight velocity at 40,000 ft and at the maximum landing weight is given by

$$C_L V^2 = \frac{2W}{\rho \sigma S} = \frac{2 \times 103{,}000}{0.002377 \times 0.247 \times 980} = 358{,}000 \, \frac{\text{ft}^2}{\text{sec}^2}$$

and is shown in Table 3.1. The table shows that, theoretically, at 40,000 ft altitude this aircraft may be able to maintain steady level

TABLE 3.1 Lift and Velocity Comparisons

V (ft/sec)	352	400	423	500	600	700	800	850
C_L	2.89	2.24	2.0	1.43	0.99	0.731	0.559	0.496

flight at 423 ft/sec at the point of stalling out the aircraft. In practice, for safety reasons, such low (stall) speed should never be attempted. Moreover, near-stall high lift coefficients may cause sufficiently high drag that can exceed the powerplant capability.

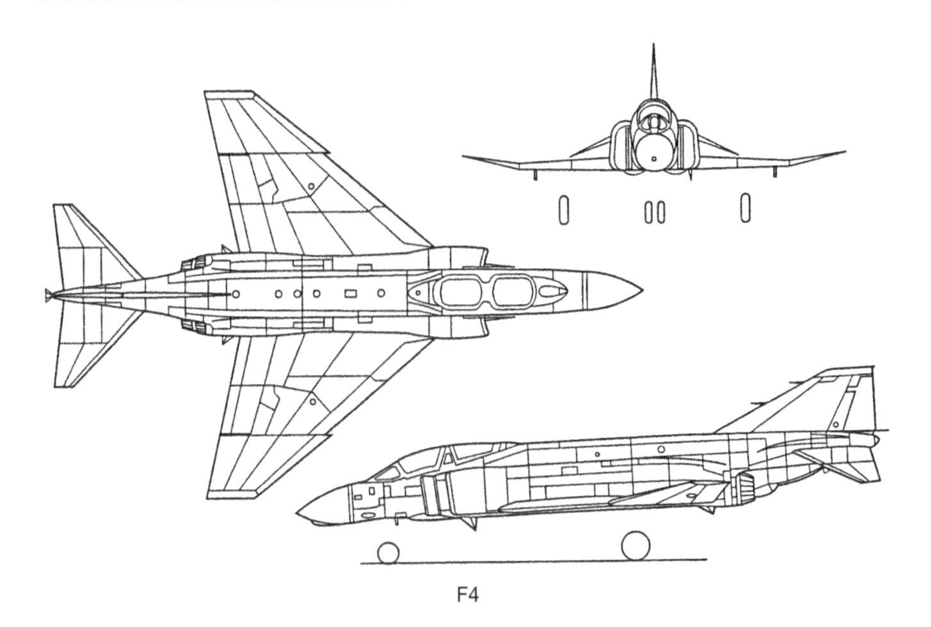

F4

3.3 MAXIMUM VELOCITY AND CEILING

3.3.1 General Considerations

In the last section, the low end of flight spectrum was explored in some detail. It was found that, for steady flight condition and in absence of concerns about thrust, there exists a minimum velocity—the stall speed. Once the aircraft minimum velocity is established, the following question naturally arises: What is the maximum speed for the aircraft, and how can it be determined? The answer to the first question is obviously affirmative, but how it is determined depends on the information available for the vehicle and is the subject of this section. Establishing the ceiling—how high can an aircraft fly, can be accomplished in a number of ways and it is discussed in several chapters. In this section it appears as a byproduct of basic thrust—drag considerations.

Consider unaccelerated flight at constant altitude. The FPE (Eq. 3.2) becomes

$$P_a = DV \equiv P_r \tag{3.13}$$

Thus, the problem is reduced to finding the flight condition where the available power just balances the energy dissipated by drag, or the power available is equal to the power required. The problem must now be further subdivided into two categories because there are basically two types of powerplants used in aircraft:

- Power producing—reciprocating engines and turbine engines driving a propeller
- Thrust producing—turbojets, ramjets, etc.

Power Producing The need for such a division becomes apparent if one considers Eq. 3.13. Fundamentally, it is a power equation where the power is thrust horsepower and is given by

$$P_a = BHP\ \eta_p \text{ in } HP \tag{3.14}$$

where BHP is the total brake horsepower produced by the engines and η_p is the propeller efficiency. Since the reciprocating and the turboprop engines are power-producing devices, the performance is described by the product of the engine shaft horsepower and the efficiency of the propeller. Thus, Eq. 3.13 may be written as

$$BHP\eta_p = DV \tag{3.15}$$

In general, both the horsepower and efficiency are functions of velocity and altitude and the power available is usually given either in graphical or tabular format (see Ex. 3.4). Table 3.2 shows typical units in use.

Thrust Producing For jet engines, the output is thrust. Since the jet power is a product of thrust and velocity, and since jet engine performance curves are usually given directly in terms thrust, Eq. 3.13 simplifies with

$$P_a = TV$$

TABLE 3.2 Units

	P	T	V
$P = TV$	$\frac{\text{ft-lb}}{\text{sec}}$	lb	$\frac{\text{ft}}{\text{sec}}$
$P = \frac{TV}{550}$	HP	lb	$\frac{\text{ft}}{\text{sec}}$
$P = TV$	W	N	$\frac{\text{m}}{\text{sec}}$

to

$$T = D \tag{3.16}$$

A comparison of Eqs. 3.15 and 3.16 shows that reciprocating engine performance should be considered from a power-available/power-required point of view and the jet engine case reduces to thrust-available/thrust-required formulation. The thrust of a jet engine can often be described by analytical statements that can lead to simple performance calculations and analytical solutions (Appendix D). For propeller-powered aircraft, the power information is usually given in graphical form, which implies that the propeller engine performance problems are best solved in graphical or numerical form. As in the case for jet aircraft, simplified solutions can be found if the aircraft is operating essentially at constant velocity and at constant altitude. In this case, propeller efficiency remains practically constant and the power available curve is also constant independent of velocity. Thus, the different formulations of thrust and power data indicate that it is advantageous to pursue the questions concerning the maximum velocity and ceiling separately for power- and thrust-oriented aircraft.

Both the P_a versus P_r and T_a versus T_r analyses presuppose a total knowledge of the aircraft drag. Thus, it will be useful to review the drag expressions in some detail. As a result, several specific relationships of general interest can be established that will be of use in simplifying the general performance calculations carried out in subsequent chapters.

747

3.3.2 Drag and Drag Polar

Although the drag and the drag coefficient can be expressed in a number of ways, for reasons of simplicity and clarity, the parabolic drag polar will be used in all main analyses. For most of the existing (high-speed) aircraft, the drag cannot be adequately described by such a simplified expression. Exact calculations must be carried out using extended equations or tabular data. However, the inclusion of more precise expressions for drag at this stage will not greatly enhance basic understanding of performance, and thus, will be included only in some calculated examples and exercises.

Writing now the drag force as

$$D = \frac{1}{2}\rho V^2 C_D S \tag{3.17}$$

and utilizing the parabolic drag polar (see Appendix B)

$$C_D = C_{D_0} + C_L^2/(\pi A\!\!R e) = C_{D_0} + kC_L^2 \tag{3.18}$$

where

$$k = 1/(\pi A\!\!R e)$$

One obtains

$$D = \frac{1}{2}\rho V^2 C_{D_0} S + \frac{1}{2}\rho V^2 S k C_L^2 \tag{3.19}$$

where

C_{D_0} is zero lift drag coefficient
$A\!\!R = b^2/S \equiv$ aspect ratio
b is wing span

Eliminating C_L from the last equation by use of Eq. 3.5, and defining the aircraft load factor n as

$$n = \frac{L}{W} \tag{3.20}$$

the drag equation becomes

$$D = \frac{1}{2}\rho V^2 C_{D_0} S + \frac{kn^2W^2}{\frac{1}{2}\rho V^2 S} \tag{3.21}$$

Eq. 3.21 is given by two terms, one proportional to V^2 and the other inversely proportional to V^2. The first term, called *parasite drag,* represents the aerodynamic cleanness with respect to frictional characteristics, and shape and protuberances such as cockpit, antennae, or external fuel tanks. It increases with the aircraft velocity and is the main factor in determining the aircraft maximum speed. The second term represents induced drag (drag due to lift). Its contribution is highest at low velocities/high-g loading, and it decreases with increasing flight velocities/lower-g flight. For the rest of this chapter and for most of Chapters 4, 5, and 6, it is assumed that $L = W$, and therefore $n = 1$.

The relative significance of these two terms of Eq. 3.21 is shown in Fig. 3.1, where a typical total drag curve is drawn for a parabolic drag polar at sea-level altitude ($\rho = \rho_0$). The effect of increase in altitude (decrease in ρ) is also shown, which shifts the total drag curves right for constant W, n, S, and C_{D_0}. Increasing W, n, S, and C_{D_0} increases the drag and curves shift upward. Thus, the drag curve depends on five parameters, out of which only the reference area S will usually remain constant during the flight. Although n is unity for steady level flight, as considered in most of these chapters, it will be included here explicitly for basic development. This will permit an easy transfer to other cases in later chapters.

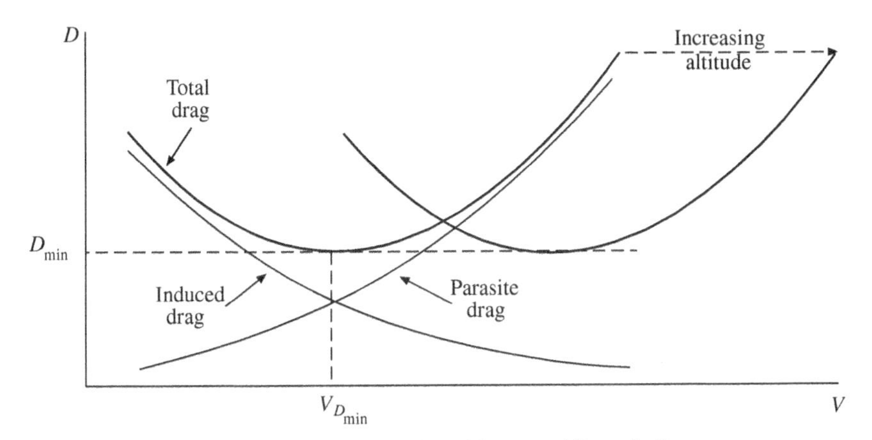

Figure 3.1 Total Drag, Induced Drag, and Parasite Drag

For low-speed flight regime (where the Mach number effects can be ignored), it is convenient to rewrite Eq. 3.21 by introducing the equivalent parasite area:

$$f = C_{D_0}S \tag{3.22}$$

$$D = \frac{\rho}{2} fV^2 + \left(\frac{W}{b}\right)^2 \frac{2n^2}{\pi e \rho V^2} \tag{3.23}$$

The equivalent parasite area is often used to classify different aircraft and their component drags, and it leads to a logical way to a buildup of aircraft zero lift drag. For some tabulated data see Appendix C.

Since a graphical performance representation is often a convenient means toward solutions, the number of apparent parameters can be reduced by suppressing the altitude variation by introducing again the equivalent air speed from Eq. 3.10, which gives

$$D = \frac{1}{2}\rho_0 V_E^2 C_{D_0} S + \frac{2kn^2W^2}{\rho_0 V_E^2 S} = AV_E^2 + \frac{B}{V_E^2} \tag{3.24}$$

where

$$A = \frac{1}{2}\rho_0 C_{D_0} S$$

$$B = \frac{2kn^2W^2}{\rho_0 S}$$

and A and B depend on geometry, weight, and g-loading of the aircraft. Thus, for a given aircraft point performance problem at fixed weight and g-load, the drag can be given by a single curve for all altitudes, which helps to simplify graphical performance representation (compare Figures 3.1 and 3.2).

An item of considerable practical interest is the minimum point of the total drag curve, as it gives the minimum drag and the maximum lift/drag ratio and later serves to identify useful performance points. To find this point on the drag curve, any one of the drag expressions given above, say Eq. 3.21, can be differentiated with respect to the velocity to give

$$\frac{dD}{dV} = \rho V C_{D_0} S - \frac{4kn^2W^2}{\rho V^3 S} \tag{3.25}$$

Setting this expression equal to zero, and solving for velocity, one finds

$$V_{D_{\min}} = \sqrt{\frac{2Wn}{\rho S}} \left(\frac{k}{C_{D_0}}\right)^{1/4} \tag{3.26}$$

Substituting Eq. 3.26 back into the drag equation Eq. 3.21, yields

$$D_{\min} = Wn\sqrt{kC_{D_0}} + Wn\sqrt{kC_{D_0}} = 2Wn\sqrt{kC_{D_0}} \tag{3.27}$$

which states that the minimum drag occurs when the parasite and induced drags are equal and it is independent of altitude. From the drag coefficient, Eq. 3.18, it follows then, since both of the drag terms are equal (this can be seen also from Fig. 3.1), that

$$C_{D_0} = kC_L^2 \tag{3.28}$$

and the total drag coefficient, at minimum drag, may be written as

$$C_D = C_{D_0} + kC_L^2 = 2C_{D_0} = 2kC_L^2 \tag{3.29}$$

Since the minimum drag occurs at the minimum value of D/L, the latter ratio can be determined immediately by use of Eqs. 3.28 and 3.29 as

$$\left.\frac{D}{L}\right|_{\min} = \frac{1}{\left.\frac{L}{D}\right|_{\max}} \equiv \frac{1}{E_m} \tag{3.30}$$

$$\left.\frac{L}{D}\right|_{\max} = \left.\frac{C_L}{C_D}\right|_{\max} \equiv E_m = \sqrt{\frac{C_{D_0}}{k}} \left(\frac{1}{2C_{D_0}}\right) = \frac{1}{2\sqrt{kC_{D_0}}} \tag{3.31}$$

NOTE

In keeping with the current practice in nomenclature usage, in the rest of the chapters the symbol E_m will be used for the maximum lift/drag ratio $L/D|_{max}$. It should be noted that, for the parabolic drag polar, the velocity for minimum drag occurs at E_m; thus $V_{D_{\min}} = V_{E_m}$ and that the minimum drag, Eq. 3.27, is independent of altitude, which proves the assertion made above: the effect of altitude change is to shift the drag curve parallel to the velocity axis. As with advanced nomencature, care should be exercised in usage (e.g., using $V_{D_{\min}} = V_{E_m}$ may be ambiguous). Does E refer to

L/D ratio or to equivalent airspeed EAS? In this case it is best to use full subscript $V_{(L/D)|\max}$, or simply $V_{D_{\min}}$. The next example explores calculation of the drag parameters discussed above.

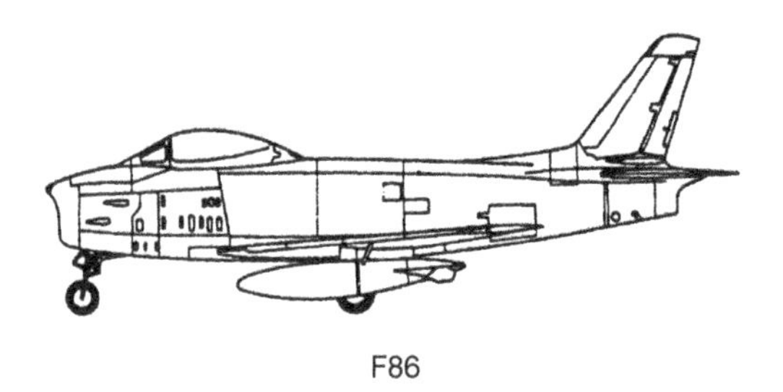

F86

EXAMPLE 3.2

An aircraft has a wing area of 255 ft^2 and a weight of 10,000 lb and a clean drag polar (flaps and gear up) of

$$C_D = 0.023 + 0.0735\ C_L^2,\ AR = 5.07$$

Calculate:

a. $(L/D)|_{\max}$
b. $V_{D_{\min}}$ at sea level and at 40,000 ft
c. $T_{\min}$ for level flight

a. Eq. 3.31 gives E_m directly

$$\left.\frac{L}{D}\right|_{\max} = \frac{1}{2\sqrt{kC_{D0}}} = \frac{1}{2\sqrt{0.0735 \times 0.023}} = 12.16$$

or, by Eq. 3.28 $C_{D_0} = kC_L^2$, we get:

$$C_{L_{D_{\min}}} = \sqrt{\frac{C_{D0}}{k}} = \sqrt{\frac{0.023}{0.0735}} = 0.559$$

and by use of Eq. 3.31, one finds

$$E_m = \left.\frac{C_L}{C_D}\right|_{\max} = \frac{C_L}{2C_{D_0}} = \frac{0.559}{2 \times 0.023} = 12.16$$

b. The velocity (EAS) at minimum drag can be found from Eq. 3.26 ($n = 1$)

$$V_{E_{D_{\min}}} = \sqrt{\frac{2W}{\rho_0 S}}\left(\frac{k}{C_{D_0}}\right)^{1/4} = \sqrt{\frac{2 \times 10{,}000}{0.002377 \times 255}}\left(\frac{0.0735}{0.023}\right)^{1/4}$$

$$= 242.8 \frac{\text{ft}}{\text{sec}}$$

This is also $V_{D_{\min}}$ (TAS) at sea level. At 40,000 feet ($\sqrt{\sigma} = 0.497$)

$$V_{D_{\min}} = \frac{242.8}{0.497} = 488.6 \frac{\text{ft}}{\text{sec}}$$

c. The minimum thrust can also be found in a number of ways:

1. From E_m, and for steady, level flight $(W/T) = (L/D)$ one finds from Eq. 3.7

$$T_{\min} = \frac{W}{E_m} = \frac{10{,}000}{12.16} = 822.4 \text{ lb}$$

which is independent of altitude.

2. $D_{\min}$ can be obtained from Eq. 3.27, which gives

$$D_{\min} = 2W\sqrt{kC_{D_0}} = 2 \times 10{,}000 \times \sqrt{0.0735 \times 0.023}$$

$$= 822.3 \text{ lb}$$

3. From the basic drag expression, which can be written by use of Eq. 3.17

$$D_{\min} = \frac{1}{2}\rho_0 V^2_{E_{D_{\min}}} C_{D_{\min}} S = \rho_0 V^2_{E_{D_{\min}}} C_{D_0} S$$

$$= \frac{1}{2} 0.002377 \times (242.8)^2 \times 2 \times 0.023 \times 255$$

$$= 821.9 \text{ lb}$$

where $$C_{D_{\min}} = 2C_{D_0} = 2 \times .023$$

Checking for $L/D|_{max}$

$$\frac{L}{D} = \frac{W}{D_{min}} = \frac{10{,}000}{821.9} = 12.17$$

a value in agreement with calculations above.

These results are found also in Figure 3.4. The calculated relationships for D_{min} and for $V_{D_{min}}$ can be read directly off the graph. Also shown are induced drag

$$D_i = \frac{2kW^2}{\rho_0 V_E^2 S}$$

and the parasite drag

$$D_p = \frac{1}{2}\rho_0 V_E^2 C_{D_0} S$$

for a weight of 10,000 lb. The influence of increased weight is also shown for 12,000 and 14,000 lb.

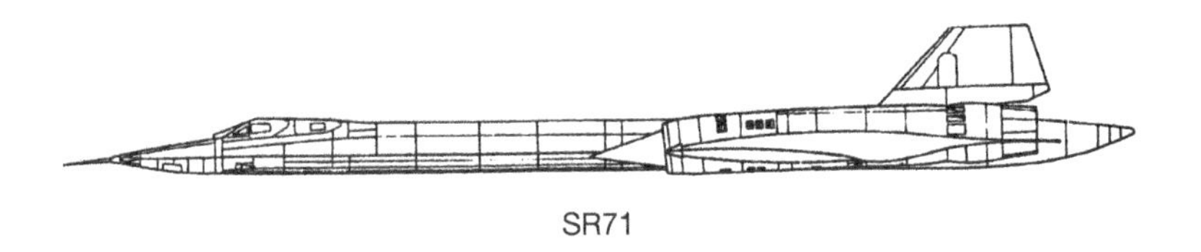

SR71

3.3.3 Flight Envelope: V_{max}, V_{min}

Returning to the original purpose of this chapter, to determine the maximum and minimum velocities, it is necessary to consider the thrust-drag or power balance of the aircraft. For jet engine operation, the power-available versus power-required statement reduces to thrust-required versus thrust-available, as already shown by Eq. 3.16 and Eq. 3.24.

$$T_a = T_r = D = \frac{1}{2}\rho V^2 C_D S = AV_E^2 + \frac{B}{V_E^2} \qquad (3.32)$$

Since, for steady, level flight, the thrust required is equal to the aircraft drag, the developments in Subsection 3.3.2 can be applied here directly, and Eqs. 3.18 to 3.31 give appropriate results for thrust required, minimum thrust, and so on as shown in Example 3.2. If the thrust available is known, then Eq. 3.32 can be solved directly for maximum or minimum velocity at a given altitude and weight (at constant values of A and B). In general, this means solving a fourth-degree algebraic equation with two positive roots providing the maximum and minimum velocities. Solution methods for such (polynomial) equations have been known for a long time, but the amount of labor and time involved has resulted in preference to the graphical approach described in the next section (the *exact* method). Currently, any advanced hand-held calculator can solve this problem in seconds (the Solver button), or one may turn to a number of computer applications, such as MATLAB.

Unfortunately, jet engine performance data are not usually available in a format that will permit a simple solution without recourse to simplifying assumptions. Thus, the method of solution of Eq. 3.32 depends on the type of engine data format—tabular, graphical, or equations—and willingness or need to use any of the appropriate simplifying equations found in Appendix D.

Two practical methods are found in general use:

1. The *exact* or complete method, where detailed engine thrust data are plotted over the drag curves on the $T - V$, (or V_E) map with altitude as a parameter (Figure 3.2). The more detailed curves may contain also fuel and air flow data (Appendix D). Exactness here depends on the completeness of drag data (i.e., inclusion of

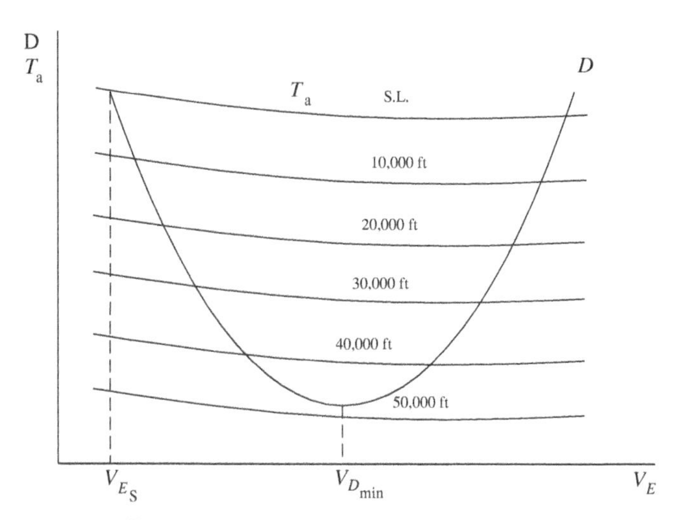

Figure 3.2 Aircraft Thrust Available and Drag

the effects of compressibility and high lift devices, stores, etc.) and the precision of drawing the curves and reading the results from the curves. The real advantage of this approach lies in the overview it provides of the aircraft overall performance potential. Thus, it is also one way of presenting the aircraft flight envelope.

2. An approximate approach where the thrust expressions of the type of Item D.1 or D.2 (Appendix D) are used to arrive at analytical solutions.

The Exact Method As shown in Figure 3.2, one finds typical subsonic jet engine data plotted over the drag data where the thrust decreases with flight velocity (or Mach number) and shows a substantial decrease with altitude. It should be noted that what is plotted is usually the maximum available thrust at a given engine rating. Any thrust between that level and idle thrust is available and is a function of the throttle setting. Moreover, standard engine curves are for uninstalled engines. For (subsonic) installed engies the data should be derated by 5 to 10 percent, depending on the aircraft and installation. Since supersonic engine data is a strong function of the intake configuration, the data is best used in installed format.

The following information is immediately available from the curves shown in Figure 3.2:

- Maximum and minimum velocities are found, at various different altitudes, at the intersection of thrust and drag curves. For a given altitude, the thrust curve crosses the drag curve at two widely different velocities yielding the maximum and minimum level flight velocities. The corresponding equivalent airspeed is read directly off the abcissa. Figure 3.3 shows typical EAS and TAS plotted against altitude. It is seen that, for jet aircraft, the maximum true airspeed occurs at some intermediate altitude between sea level and the ceiling. Since the overall V_{max} occurs at an altitude higher than sea level, the thrust required has also been reduced from that at the sea level (see Figure 3.3). As the jet engine fuel consumption is proportional to the thrust, it follows immediately that more economical flight occurs at altitudes higher than sea level.

  It is seen that, for a given thrust level (altitude), there is a single V_{max}. It should be noted that there are typically two minimum velocities: the stall velocity V_s (Section 3.2) and a minimum value V_{min} as determined by the available thrust level. If V_s is greater

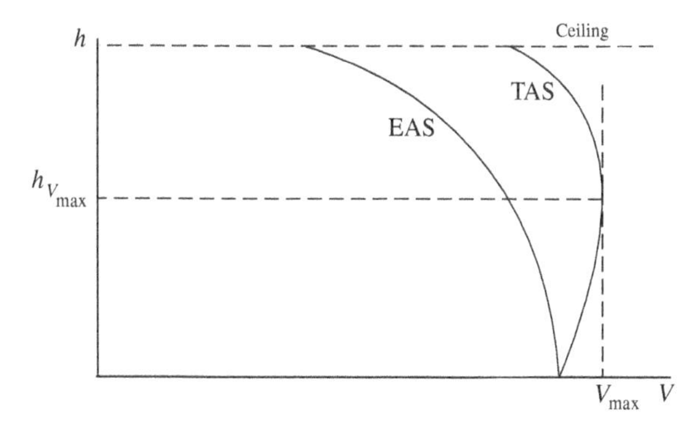

Figure 3.3 Airspeed as a Function of Altitude

than V_{min} then, for steady level flight, the aircraft minimum velocity is determined by the maximum lift available at V_s. The relative magnitude of these two velocities is determined by the aircraft maximum lift coefficient and the characteristics of the particular powerplant used.

- Ceiling also can be found directly from Figure 3.2. For steady, level flight, ceiling is defined by the condition where, at highest altitude, $T_a = D$. This means, that the (absolute) ceiling is found at the locus of the highest altitude T_a curve being tangent to a D (T_{req}) curve. The tangency condition also determines the velocity at which the (absolute) ceiling may be reached. For complete ceiling definitions, see the next chapter.
- The slight negative slope of the thrust curves in Figure 3.2 indicates that the velocity at the ceiling is at or near $V_{D_{min}}$. In this case, the ceiling is shown to be at about 48,000 ft. If the thrust available curves were independent of altitude (a straight horizontal line), then the velocity at ceiling would occur at $V_{D_{min}}$ and then also it is seen that at ceiling $V_{D_{min}} = V_{min} = V_{max}$.

The results and graphs shown in Figures 3.2 and 3.4 require repetitious calculations for a given aircraft configuration and required data of the thrust as a function of velocity and altitude. However, the amount of extra labor provides a general overview and realistic results with good precision. Example 3.3a shows the full calculation procedure to establish the flight envelope for the aircraft discussed in Example 3.2.

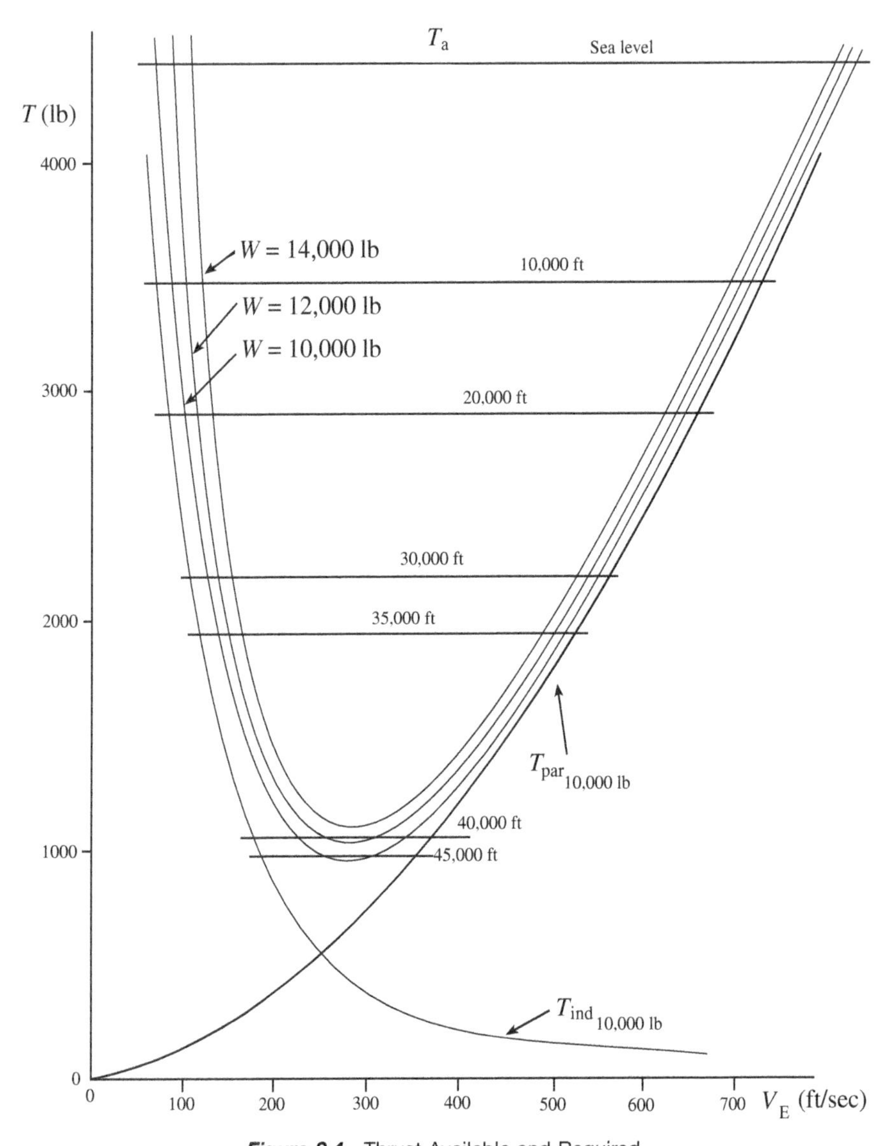

Figure 3.4 Thrust Available and Required

EXAMPLE 3.3a

Suppose that the example aircraft treated in Example 3.2 is equipped with two jet engines of P&W-60 class. The aircraft characteristics are as follows:

$$C_D = 0.023 + 0.0735\ C_L^2$$

$$S = 255\ \text{ft}^2$$

$$W = 12{,}000\ \text{lb}$$

$$C_{L_m} = 1.8$$

$$T = T_o\sigma^m \text{ per engine}$$

where $T_o = 2{,}200$ lb

$$m = 0.7,\ h < 36{,}000\ \text{ft}$$

$$m = 1,\quad h > 36{,}000\ \text{ft}$$

The engine data is somewhat simplified (see Appendix D), and is taken to be independent of flight velocity. The drag has been calculated from Eq. 3.24 with

$$A = \frac{1}{2}.002377 \times .023 \times 255 = .00697$$

$$B = 2 \times .0735 \times W^2/.002377/255 = .2425W^2$$

yielding

$$D = .00696V_E^2 + .2425W^2/V_E^2$$

The results, with engine data for twin engines, are shown in Figure 3.4 for the aircraft weight ranging from 10,000 to 14,000 lb. Also shown are the individual curves for the induced and parasite drags for a weight of 10,000 lb. As expected, these drag terms are equal at $V_{ED_{min}} = 242.8$ ft/sec (see Example 3.2).

The maximum speed at 30,000 ft altitude is read off as

$$V_{max30} = V_{E_{max30}}/\sqrt{\sigma_{30}} = 547/\sqrt{.375} = 893\ \text{ft/sec}$$

Also, the ceiling can be read off directly where the (imaginary) thrust line is tangent to 12,000 lb drag curve—at about 41,000 ft. Determining the other results such as rate of climb, a more precise ceiling, and so on, will be discussed later in appropriate chapters.

The Aproximate Approach This method provides quick results with less labor but also with a possible loss of accuracy. Here, drag data is given by the drag polar, and the (jet) engine data is approximated, usually by one of the analytical expressions found in Appendix D. For discussion sake and to present the methodology jet engines, the data leading to Figure 3.2 can be obtained approximately by one of the following expressions:

$$T = T_0 \sigma^m \tag{3.33}$$

$$T = (A + BV^2)\sigma^m \tag{3.34}$$

where A, B, and m are constants obtained from curve-fitting the manufacturer-provided engine data and are not to be confused with the drag representation with Eq. 3.24. Eqs. 3.33 and 3.34 lead to analytic expressions for aircraft performance, with accuracy depending on the goodness of fit to actual engine data.

To simplify calculations, and to generalize the results, it is practical to use nondimensionalized equations, starting with Eq. 3.21 and introducing $V_{D_{\min}}$ as a reference condition at steady, level flight ($n = 1$), and a nondimensional velocity $\overline{V}$ as follows:

$$\overline{V} = \frac{V}{V_{D_{\min}}}$$

$$D = \frac{1}{2}\rho C_{D_0} V^2 S \left(\frac{2W}{S}\right)\sqrt{\frac{k}{C_{D_0}}} + \frac{kn^2W^2}{\frac{1}{2}\rho S\left(\frac{2W}{\rho S}\right)\sqrt{\frac{k}{C_{D_0}}}\,\overline{V}^2} \tag{3.35}$$

Since $(L/D)_{\max} \equiv E_{\rm m} = 1/(2\sqrt{kC_{D_0}})$, this drag equation can be rewritten as

$$\frac{D}{W} = \frac{1}{2E_{\rm m}}\left(\overline{V}^2 + \frac{n^2}{\overline{V}^2}\right) \tag{3.36}$$

By differentiating with respect to nondimensional velocity $\overline{V}$, one can recover the results already obtained in Subsection 3.3.2. The minimum drag occurs at

$$\overline{V} = \sqrt{n} \tag{3.37}$$

and is given by (W = constant)

$$\left.\frac{D}{W}\right|_{\min} = \frac{n}{L/D|_{\max}} \tag{3.38}$$

For steady, level flight, $n = 1$ and these results simplify further to

$$\overline{V} = 1 \tag{3.39}$$

$$\left.\frac{D}{W}\right|_{\min} = \frac{1}{L/D|_{\max}} = \frac{1}{E_{\mathrm{m}}} \tag{3.40}$$

which is Eq. 3.30 and was entirely expected. The maximum velocity can now be obtained by substituting a suitable thrust expression into Eq. 3.36.

Consider first Eq. 3.33—or for that matter, any thrust expression with T independent of velocity. Then Eq. 3.36 may be written, for steady, level flight:

$$\frac{2TE_{\mathrm{m}}}{W} - \left(\overline{V}^2 + \frac{1}{\overline{V}^2}\right) = 0 \tag{3.41}$$

Introducing now a dimensionless thrust $\overline{T}$

$$\overline{T} = \frac{TE_{\mathrm{m}}}{W} \tag{3.42}$$

one obtains from Eq. 3.41

$$\overline{V}^4 - 2\overline{T}\overline{V}^2 + 1 = 0 \tag{3.43}$$

Eq. 3.43 can be solved to give

$$\overline{V}^2 = \overline{T} \pm \sqrt{\overline{T}^2 - 1} \tag{3.44}$$

which, in turn, yields two more solutions. Only the two following positive solutions are physically meaningful:

$$\overline{V}_1 = \sqrt{\overline{T} - \sqrt{\overline{T}^2 - 1}} \tag{3.45}$$

$$\overline{V}_2 = \sqrt{\overline{T} + \sqrt{\overline{T}^2 - 1}} \tag{3.46}$$

It is evident that $\overline{V}_2$ represents the high-speed solution, $\overline{V}_2 > 1$, and $\overline{V}_1$ is the low-speed solution, $\overline{V}_1 < 1$. This follows from the fact that, for real solutions, $\overline{T}$ must be larger than unity. If $\overline{T} = 1$, one sees immediately that

$$\overline{V}_1 = \overline{V}_2 = 1 \tag{3.47}$$

which represents the condition where minimum and maximum velocities coincide, and implies that $V = V_{D_{min}}$. Clearly, then, this gives a solution for the ceiling where $\overline{T} = 1$ and the velocity at the ceiling $V_{ceiling} = V_{D_{min}}$. Incidentally, Eqs. 3.45 and 3.46 yield, after multiplication by each other, an interesting condition: $\overline{V}_1\overline{V}_2 = 1$. This restates the conclusion already obtained that one of the nondimensional velocities must be larger than, and the other smaller than, unity. Thus, if one solution is known, the other can be easily found from this condition.

The ceiling can now be calculated from the condition that $\overline{T} = 1$:

$$\overline{T} \equiv \frac{TE_m}{W} = 1 \tag{3.48}$$

Substituting for the thrust from Eq. 3.33, one gets the density ratio at the ceiling:

$$\sigma_c = \left(\frac{W}{T_o E_m}\right)^{1/m} = \left(\frac{2W\sqrt{kC_{D0}}}{T_o}\right)^{1/m} \tag{3.49}$$

Caution should be exercised in applying Eq. 3.49, since the exponent m in the thrust equation may not apply over the entire altitude range (see Example 3.3a and Appendix D).

The maximum velocity can be calculated from $\overline{V}_2$. Substituting $\overline{T}$ and recalling the definition of $\overline{V}$, one obtains from Eq. 3.46

$$\frac{V_{max}}{V_{D_{min}}} = \sqrt{\frac{TE_m}{W} + \sqrt{\left(\frac{TE_m}{W}\right)^2 - 1}} \tag{3.50}$$

Eq. 3.50 gives the maximum velocity, at a given altitude, for given T, W, S, k, C_D. However, the altitude at which the maximum aircraft velocity occurs cannot be found easily from Eq. 3.50 (see Figure 3.3). To simplify calculations, nondimensional velocity $\overline{V}$ shall be redefined in terms of the sea level $V_{D_{min0}} \equiv V_{E_{D_{min}}}$ as follows by use of Eq. 3.26:

$$\overline{V}_{\mathrm{o}} = \frac{V}{V_{D_{\mathrm{min_o}}}} = \frac{V}{\sqrt{\dfrac{2W}{\rho_0 S}\left(\dfrac{k}{C_{D_0}}\right)^{1/4}}} \tag{3.51}$$

Introducing this into Eq. 3.50 and using Eq. 3.33, one obtains, after some rearrangement (dividing both sides by $\sqrt{\sigma}$),

$$\overline{V}_{\mathrm{o_{max}}} = \sqrt{\frac{T_{\mathrm{o}}E_{\mathrm{m}}}{W\sigma^{1-m}} + \sqrt{\left(\frac{T_{\mathrm{o}}E_{\mathrm{m}}}{W\sigma^{1-m}}\right)^2 - \frac{1}{\sigma^2}}} \tag{3.52}$$

where T_{o} is the sea-level engine thrust and E_{m} is given by Eq. 3.31. It is seen that the altitude for V_{max} increases as the thrust level is increased.

To round off the approximate method, a second approach to the velocity solution is to set the thrust available equal to thrust required, $T_{\mathrm{a}} = T_{\mathrm{r}} = D$, by use of Eq. 3.21, as follows:

$$T_{\mathrm{a}} = \frac{1}{2}\rho V^2 S C_{D_0} + \frac{2kW^2}{\rho V^2 S} \tag{3.53}$$

and to solve for the velocity V. After some laborious manipulations, one obtains

$$V = \sqrt{\frac{T_{\mathrm{a}}}{\rho S C_{D_0}}\left[1 \pm \sqrt{1 - \frac{1}{(E_{\mathrm{m}}(T_{\mathrm{a}}/W))^2}}\right]} \tag{3.54}$$

where the positive sign is used for maximum velocity and negative sign for minimum velocity. It should be noted that T_{a} represents any available thrust (at a given throttle setting) at an altitude and is assumed to be independent of velocity in Eqs. 3.53 and 3.54. Thus, Eq. 3.54 avoids the curve-fitting process to determine the constants m, T_{o}, A, and B in Eqs. 3.33 or 3.34 and may seem to yield somewhat more accurate results. However, if thrust shows appreciable variation with velocity as is more often the case, the calculation process may turn out to be rather laborious and iterative to determine the correct value of T_{a} for use in Eq. 3.54.

EXAMPLE 3.3b

(Example 3.3a continued) This example shows how the same performance data can be determined by the approximate approach. However, since simplified engine data is used, the agreement, as anticipated, is very good. In practice, such agreement should always not be expected.

In this example, V_{max} and the ceiling are to be determined. Since E_m has been already calculated in Example 3.2 as 12.16, then

$$T_{min} = \frac{W}{E_m} = \frac{12{,}000}{12.16} = 987 \text{ lb}$$

$V_{ED_{min}}$ is obtained from Eq. 3.26:

$$V_{ED_{min}} = \sqrt{\frac{2W}{\rho_0 S}}\left(\frac{k}{C_{D_0}}\right)^{1/4} = \sqrt{\frac{2 \times 12{,}000}{0.002377 \times 255}}\left(\frac{0.0735}{0.023}\right)^{1/4}$$

$$= 266 \frac{\text{ft}}{\text{sec}}$$

The values for other altitudes are obtained from

$$V_{D_{min}} = \frac{V_{ED_{min}}}{\sqrt{\sigma}} \ (TAS)$$

The maximum velocity at altitude is found from Eq. 3.50 or 3.52. Using Eq. 3.42:

$$\frac{T_o E_m}{W} = \frac{4400 \times 12.16}{12{,}000} \ 4.45$$

one finds that at 30,000 ft, $\sigma = 0.375$ and

$$\overline{V}_{o_{max}} = \sqrt{\frac{4.45}{0.375^{0.3}} + \sqrt{\left(\frac{4.45}{0.375^{0.3}}\right)^2 - \frac{1}{0.375^2}}} = 3.36$$

Whereupon

$$V_{max_{30,000}} = 3.36\ V_{E_{D_{min}}} = 3.36 \times 266 = 894\ \frac{ft}{sec}$$

which agrees well with 893 ft/sec found in Example 3.3a. Similarly, $V_{E_{min}}$ can be obtained by first calculating from Eq. 3.45

$$\overline{V}_{o_{min}} = \sqrt{\frac{4.45}{.375^{.3}} - \sqrt{\left(\frac{4.45}{.375^{.3}}\right)^2 - \frac{1}{.375^2}}} = .7926$$

and then

$$V_{min_{30,000}} = .7926 V_{E_{D_{min}}} = .7926 \times 266 = 210\ \frac{ft}{sec}$$

The maximum and minimum velocities at each altitude are shown in Table 3.3. To find the altitude at which the aircraft achieves its absolute maximum velocity, it is easiest to plot the maximum values as function of altitude, and to determine the maximum value from that curve. A rough interpolation from Table 3.3 indicates that the maximum velocity is about 900 ft/sec, and it occurs near 35,000 ft altitude.

The density ratio at ceiling is obtained from Eq. 3.49:

$$\sigma_c = \left(\frac{W}{T_o E_m}\right)^{1/m} = \frac{1}{4.45} = 0.225$$

which gives, from the altitude table, $h = 42{,}000$ ft. The exponent, m has been assumed to be unity, as the ceiling was expected to occur at an altitude higher than 36,000 ft. In a later section in Chapter 4, where the rate of climb is studied, the ceiling can be determined more accurately since the ceiling can also be defined as the location where the rate of climb is zero.

TABLE 3.3 Velocity Comparisons, ft/sec; W = 12,000 lb

h (ft)	0	10,000	20,000	30,000	40,000	42,000
$V_{D_{min}}$	266	310	364	435	535	561
V_{max}	790	826	862	891	684	561
$V_{E_{max}}$	790	710	629	546	334	266
V_{min}	90	100	155	210	430	561

For completeness, the stall velocity is obtained from

$$V_{E_s} = \sqrt{\frac{2W}{\rho_0 S C_{L_m}}} = \sqrt{\frac{2 \times 12{,}000}{.002377 \times 255 \times 1.8}} = 148 \ \frac{\text{ft}}{\text{sec}}$$

The last line shows $V_{E_{max}}$ for direct comparison with the values obtained from Figure 3.5, where T_a and D are plotted as functions of V_E.

The results obtained so far in this chapter and in Chapter 2 can be summarized and collected in Figure 3.5, which is commonly called the *flight envelope.* At this point, the graph of V_{min} and V_{max} represents, at a given thrust rating, only the theoretical steady, level flight boundary of an aircraft. It has been calculated by use of a simple drag polar without any practical regards concerning the aircraft lifting capability (Table 3.3). Usually, V_s is larger than the drag polar calculated V_{min} over most of the lower speed range, thus shrinking the aircraft performance range. The resulting boundary shows, to a first approximation, the possible velocity range at a given altitude and also the ceiling. In conjunction with Figure 3.4 it also determines the (excess) energy available ($T_a > D$) for steady-state maneuvering within that envelope. (More discussion follows in Chapter 4, and see also Figures 4.5 and

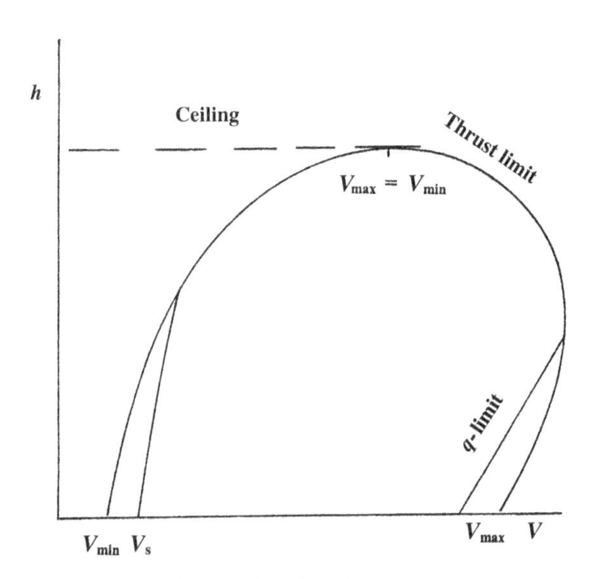

Figure 3.5 Flight Envelope

4.6.) Unsteady dynamic performance such as zooms and dives can take the aircraft for a short time outside of this envelope.

Additional realistic constraints (compressibility and planform effects on drag polar, buffeting limits, and structural and q-load considerations) may considerably decrease the V_{max} boundary. Inclusion of those effects is a must for design and realistic performance analysis but is outside the intended scope of this text. However, the methodology already outlined is applicable with any desired degree of precision and imagination.

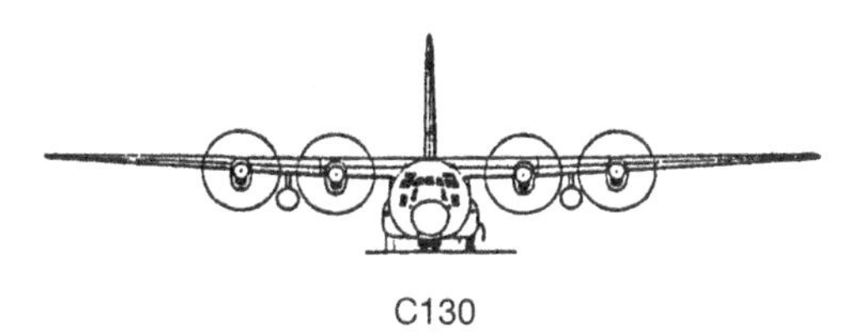

C130

3.3.4 Power Required and Power Available

It has been noted that it is more convenient to formulate propeller-driven aircraft performance in terms of power rather than via thrust expressions. The power required P_r to fly an aircraft can be found from the drag expression by use of Eqs. 3.13 and 3.19:

$$P_r = \frac{DV}{550} = \frac{\rho V^3 S C_{D0}}{1100} + \frac{k n^2 W^2}{275 \rho V S} \tag{3.55}$$

The first term gives the power required due to parasite drag. The second term is the induced power required. A typical power required curve is shown in Figure 3.6.

Due to proportionalities to V^3 and V^{-1}, the curve is skewed from the thrust required curve. The minimum power does not occur at the velocity for E_m, and the minimum power shows a strong altitude effect. To determine these respective locations, Eq. 3.55 will be differentiated with respect to the velocity. Setting the result equal to zero, one obtains for the velocity at minimum power

$$V_{P_{min}} = \sqrt{\frac{2Wn}{\rho S}} \left(\frac{k}{3C_{D0}} \right)^{1/4} \tag{3.56}$$

By comparison with Eq. 3.26, it follows immediately that

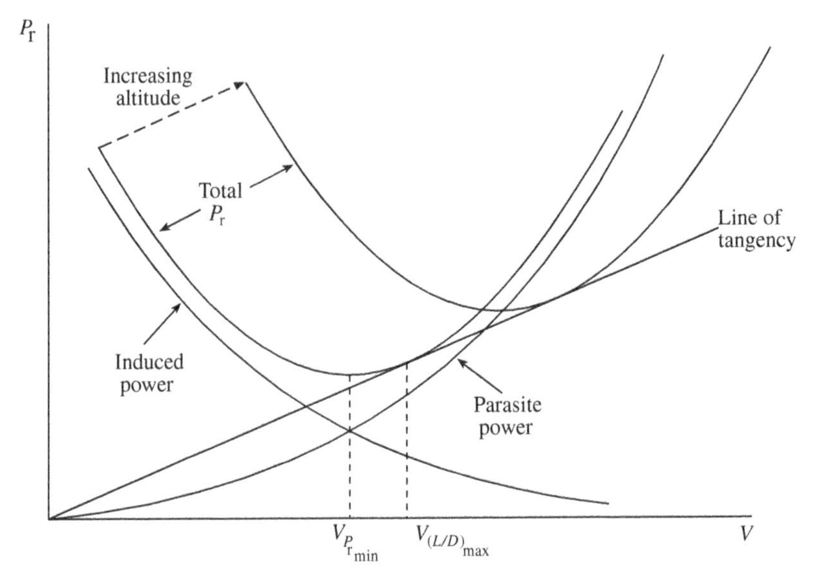

Figure 3.6 Power Required

$$V_{D_{min}} = V_{(L/D)_{max}} = 3^{1/4} V_{P_{min}} = 1.316 V_{P_{min}} \tag{3.57}$$

The minimum power required is found by substituting $V_{P_{min}}$ into Eq. 3.55, which gives for minimum power

$$\begin{aligned} P_{min} &= \frac{1}{1100} \frac{(2Wn)^{3/2}}{\sqrt{\rho S}} \left(\frac{k}{3}\right)^{3/4} C_{D_0}^{1/4} + \frac{k^{3/4}}{1100} C_{D_0}^{1/4} \frac{(2Wn)^{3/2}\, 3^{1/4}}{\sqrt{\rho S}} \\ &= \frac{2.48}{550} \frac{(Wn)^{3/2} k^{3/4} C_{D_0}^{1/4}}{\sqrt{\rho S}} \end{aligned} \tag{3.58}$$

An inspection of Eq. 3.58 reveals that, for minimum power required, the induced power required is three times the parasite power required. This implies that, at minimum power

$$kC_L^2 = 3C_{D_0} \tag{3.59}$$

$$C_D = C_{D_0} + kC_L^2 = C_{D_0} + 3C_{D_0} = 4C_{D_0} \tag{3.60}$$

The minimum power can be simplified as

$$P_{\min} = \frac{DV}{550} = \frac{1}{550}\frac{1}{2}\rho V^2 4C_{D_0}SV = \frac{C_{D_0}}{275} V^3_{P_{\min}} S\rho \qquad (3.61)$$

where $V_{P_{\min}}$ is given by Eq. 3.56. Although these minimum expressions will not be used immediately, they are useful for determining the general shape of the power required curve and are essential later for establishing other performance parameters.

Returning now to the basic problem at hand: Finding the maximum velocity (a solution to Eq. 3.13) can be easily accomplished by plotting both P_a and P_r curves versus V, as shown in Figure 3.7. The power-required curves are given for several altitudes at constant weight or nW. Power-available curves for engines as a function of altitude and RPM are supplied by the manufacturer as BHP (brake horsepower) in graphical form (see Appendix D). Power available is calculated from

$$P_a = BHP\ \eta_p \qquad (3.62)$$

Since each propeller has its own different efficiency curves (depending on its size, velocity, rpm, twist, etc.), families of P_a curves can be developed with engine rpm as a parameter. It is impractical to derive this information in analytical form. Thus, a graphical approach is used for general propeller aircraft performance calculations. For constant-altitude flight, the power-available curve may be approximated by a straight line (see Problem 3.11). Example 3.4 shows a typical approach to this problem.

EXAMPLE 3.4

Consider a twin-engine aircraft with the following characteristics:

$S = 300\ \text{ft}^2$

$AR = 7$

$e = 0.85$

$C_D = 0.024 + 0.0535\ C_L^2$

$W_{T.O.} = 9{,}600\ \text{lb}$

$W_{fuel} = 1{,}450\ \text{lb}$

The engines are Lycoming I0-720 400 HP rated at 2,650 RPM full throttle. The values for available power at 2,400 RPM are shown in Table 3.4 for altitude variation: The propellers are three-bladed Hamilton Standard, 6 ft diameter, 100 activity factor, 0.3 integrated design lift coefficient. Table 3.5 shows propeller efficiency at these altitudes as a function of flight speed. The power required has been calculated from Eq. 3.55 and is shown in Figure 3.7.

The power available was calculated by use of Eq. 3.62, BHP, and efficiency given in Tables 3.4 and 3.5. Using both Figure 3.7 and the analytical equations, the following will be determined for a weight of 9,117 lb:

a. P_{min} at sea level
b. $V_{(L/D)\text{max}}$ at sea level
c. V_{max}
d. ceiling

The answers follow in the same sequence.

a. Graphical solution is obtained from Figure 3.7 showing $P_{\text{min}} =$ 203 HP at about 146 ft/sec. It can be found also by use of Eq. 3.58 as

$$P_{\text{min}} = \frac{2.4816}{550} \frac{(9117)^{3/2}(0.0535)^{3/4}\ (0.024)^{1/4}}{\sqrt{0.002377 \times 300}} = 203 \text{ HP}$$

The velocity at minimum power follows from Eq. 3.56

$$V_{P_{\text{min}}} = \sqrt{\frac{2 \times 9117}{0.002377 \times 300}} \left(\frac{0.0535}{3 \times 0.024}\right)^{1/4} = 148 \frac{\text{ft}}{\text{sec}}$$

b. From Eq. 3.57 one finds

$$V_{(L/D)\text{max}} = 1.316 V_{P_{\text{min}}} = 1.316 \times 148 = 195 \frac{\text{ft}}{\text{sec}}$$

TABLE 3.4 Example 3.4 Engine Data

h (ft)	0	5,000	10,000	20,000
BHP	375	316	264	176

TABLE 3.5 Propeller Efficiency, η

h (ft) / *V* (mph)	0	10,000	20,000
50	0.29	0.30	
100	0.55	0.57	0.60
150	0.74	0.75	0.78
200	0.835	0.84	0.86
250	0.881	0.89	0.88

c. The values for the maximum velocity can be found from intersections of P_a and P_r curves, shown in Table 3.6.

d. The steady, level flight ceiling is given by the location where the P_r and the last P_a curves are tangent to each other. In this case, the last power available curve falls slightly below the P_r curve at 20,000 ft altitude. Interpolation of calculated values indicates that the last $P_a = P_r$ condition would occur near 19,500 ft and at about 254 ft/sec. In a later section, where the rate of climb is studied, the ceiling can be determined more accurately because the ceiling can also be defined as the location where the rate of climb is zero.

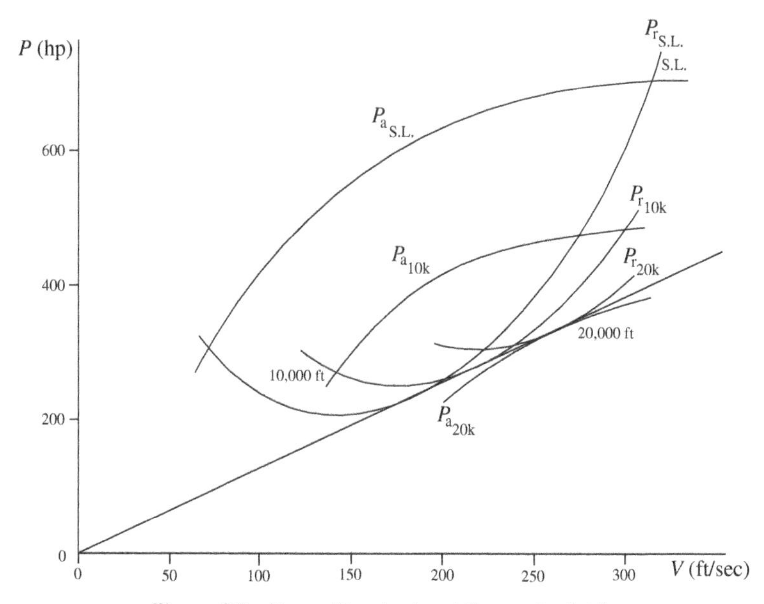

Figure 3.7 Power Required and Power Available

TABLE 3.6 Maximum Velocities

h (ft)	0	10,000	20,000
V_{max} $\left(\frac{ft}{sec}\right)$	330	300	255

Similar to the thrust-oriented approach, the intersections of the power available and power-required curves represent (the velocity) solutions to Eq. 3.55, for a given W. The areas between the P_r and P_a curves are called the steady flight envelope. In level flight, $n = 1$, and the sequence of intersections shows that there are two possible speeds for sustained level flight, V_{max} and V_{min}. From Example 3.4, the following conclusions can be drawn for propeller-driven aircraft:

1. The maximum velocity occurs at the sea level.
2. The ceiling is found from the condition where P_a and P_r curves are tangent to each other and it occurs at only one velocity—usually near $V_{(L/D)max}$ for that altitude.

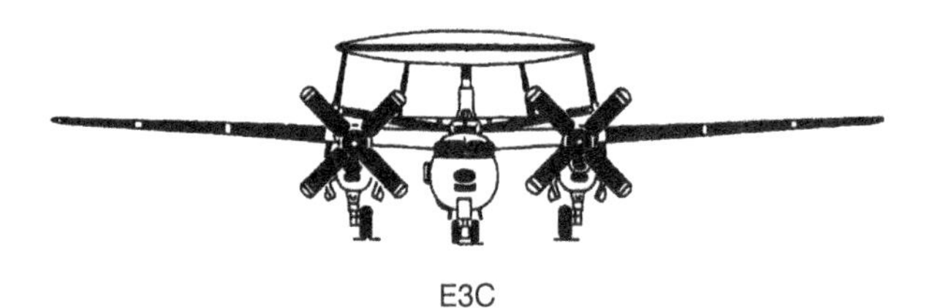

E3C

3.3.5 Turboprop Engines

A turboprop engine develops both jet and propeller thrust. A typical turboprop derives about 80 to 85 percent of the power from the shaft-propeller combination. The remaining fraction comes from the jet thrust (see Appendix D for T-56 engine data). Thus, according to the preceding sections, turboprop aircraft performance could be developed either in terms of power or in terms of thrust. Since the calculations center mainly around the propeller efficiency determination, it is convenient to analyze the turboprop performance in terms of the power expressions. The usual practice is to evaluate the thrust horsepower of a turboprop engine by means of

$$THP = SHP\ \eta_p + \frac{TV}{550} \tag{3.63}$$

where *SHP* is the installed shaft horsepower. If the velocity is given in knots, then the thrust horsepower available becomes

$$THP = SHP\ \eta_p + \frac{TV}{326} \tag{3.64}$$

Sometimes it is convenient (mainly for comparing different engines) to express the power in terms of equivalent shaft horsepower as

$$ESHP = SHP + \frac{TV}{550\eta_p} \tag{3.65}$$

from which one can calculate the total power available as

$$THP = ESHP\ \eta_p$$

The drag of a turboprop-driven aircraft must be stated in terms of power required, and the basic performance analysis follows that of a typical propeller-driven aircraft.

3.4 GLIDING FLIGHT

3.4.1 Glide Angle and Sinking Speed

There are two basic approaches for the determination of the aircraft glide performance: power-off, and partial power-on. Since the power-on glide can be viewed as a special case of Section 3.3, with a certain deficit in the thrust or power available, its discussion will be delayed until the next chapter where the excess power (or climb) problem will be studied. The foundation of all glide problems is the FPE (Eq. 3.4):

$$\frac{dh}{dt} = \frac{P_a - DV}{W} \frac{1}{\left[1 + \frac{V}{g}\frac{dV}{dh}\right]}$$

with the implied assumption that $\dot{\gamma} = 0$. This implies in Eq. 2.12 that the change of the curvature of the flight path and the associated inertia term are negligible and the energy equation Eq. 2.20 can be used directly. Since only the power-off case will be considered, Eq. 3.4 becomes

$$\frac{dh}{dt} = -\frac{DV}{W}\frac{1}{\left[1 + \frac{V}{g}\frac{dV}{dh}\right]} \tag{3.66}$$

As an energy equation, Eq. 3.66 states that the rate of change of altitude is controlled by the drag power and by the rate of change of kinetic energy (the term in the denominator). As the aircraft loses altitude, the drag power appears as the loss of potential energy, which is expended in overcoming the drag. Ignoring the kinetic energy change, the loss of potential energy and also the rate of change of altitude (sinking speed) are minimum when the drag power DV is a minimum. In glide analysis of typical slow-speed gliders and other aircraft operating in glide phase at modest altitudes, the kinetic energy change can be ignored because its contribution is small. In high-altitude operations where the flight speeds are very high (i.e., reentry vehicles), the kinetic energy change provides the major contribution to the glide range and endurance (see Example 3.6).

Following first the classical slow speed and shallow angle approach, the glide equation becomes

$$\frac{dh}{dt} = -\frac{DV}{W} \tag{3.67}$$

and the glide angle is obtained by substituting the kinematic relationship ($[dh/dt] = V \sin \gamma$) into Eq. 3.67, which gives

$$\sin \gamma \approx \gamma = -\frac{D}{W} = -\frac{C_D}{C_L} \tag{3.68}$$

with the assumption that the equilibrium conditions hold for the glide (i.e., $n = 1$, or $L = W$). It follows that the minimum glide angle γ occurs at E_{m}, and one gets

$$\gamma_{\min} = -\frac{1}{E_{\mathrm{m}}} = -2\sqrt{k\,C_{D0}} \tag{3.69}$$

The flight speed for the minimum glide angle follows (see Eqs. 3.26 and 3.31) from

$$V_{\gamma\min} = V_{(L/D)\max} = \sqrt{\frac{2W}{\rho S}}\left(\frac{k}{C_{D0}}\right)^{1/4} = V_{D\min} \tag{3.70}$$

The sinking speed $v = -dh/dt$ can be developed from Eq. 3.67 with the velocity

$$V = \sqrt{\frac{2W}{\rho S C_L}}$$

as follows:

$$v = \frac{C_D}{C_L}\sqrt{\frac{2W}{\rho S C_L}} = \sqrt{\frac{2W}{\rho S}}\,\frac{C_D}{C_L^{3/2}} \tag{3.71}$$

To find the minimum sinking speed, or the lift coefficient that minimizes the sinking speed, the above can be differentiated with respect to C_L, and the result set equal to zero. Another approach yields the same result if one recognizes, from Eq. 3.67, that v is minimum when DV (the power required) is minimum. But this condition has already been developed for Eqs. 3.56 through 3.58. As a result, one finds that the minimum sink rate $v_{\mathrm{c}\min}$ occurs at

$$V_{P\min} = \frac{V_{(L/D)\max}}{3^{1/4}} = 0.76 V_{(L/D)\max} \tag{3.72}$$

where also

$$C_D = 4C_{D0},\; k\,C_L^2 = 3C_{D0} \tag{3.73}$$

Evaluating $(L/D)_{P\min}$, one finds

$$\left.\frac{L}{D}\right|_{P_{\min}} = \left.\frac{C_L}{C_D}\right|_{P_{\min}} = \sqrt{\frac{3C_{D0}}{k}}\left(\frac{1}{4C_{D0}}\right) = \frac{\sqrt{3}}{4\sqrt{kC_{D0}}} = \frac{\sqrt{3}}{2}E_{\mathrm{m}} \quad (3.74)$$

With the aid of $V_{P_{\min}}$ and the above expression, the equation for the minimum sinking speed becomes

$$v_{c_{\min}} = \left.\frac{DV}{W}\right|_{P_{\min}} = \frac{V_{P_{\min}}}{\left.\frac{L}{D}\right|_{P_{\min}}} = 2.48\sqrt{\frac{W}{\rho S}}\,(k)^{3/4}(C_{D0})^{1/4} \quad (3.75)$$

These results can be summarized as follows:

1. The horizontal velocity $V_{P_{\min}}$ for minimum sinking speed $v_{c_{\min}}$ is smaller than the velocity for the flattest glide $V_{(L/D)_{\max}}$.
2. Both of these velocities are functions of the altitude, and they decrease as the aircraft descends toward the sea level. This is true also for the minimum sinking speed.
3. The L/D ratio is maximum for the flattest glide, but is $\sqrt{3}/2 = 0.87\,E_{\mathrm{m}}$ for the minimum sinking speed.
4. The minimum glide angle is independent of the altitude, and is only a function of E_{m}.

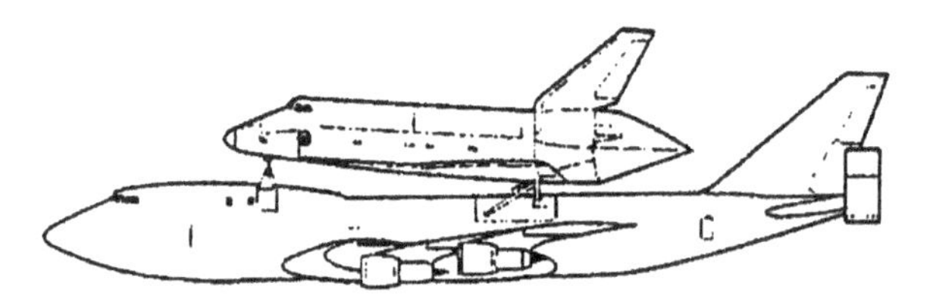

747 and Shuttle

EXAMPLE 3.5

Consider again the aircraft studied in Example 3.2. Assume that it has a total power failure at 10,000 feet. Its clean drag polar and its maximum lift to drag ratio are as follows:

$$C_D = 0.023 + 0.0735\,C_L^2$$

$$E_{\mathrm{m}} = 12.16$$

Its minimum glide angle is

$$\gamma_{\min} = \frac{1}{E_{\mathrm{m}}} = \frac{1}{12.16} = 0.0823\ rad = 4.71^\circ$$

and occurs at

$$V_{D_{\min}} = \frac{V_{E_{D_{\min}}}}{\sqrt{\sigma}} = \frac{243}{0.859} = 283\ \frac{\text{ft}}{\text{sec}}\ \text{(horizontal velocity)}$$

Assuming that the pilot can maintain an indicated airspeed of 243 ft/sec, the plane can glide a distance of

$$x = \frac{10{,}000}{0.0823} = 121{,}500\ \text{ft} = 23\ \text{miles}$$

The vertical velocity (sink speed) is Eq. 3.71 and is developed for C_D and C_L at $V_{D_{\min}}$.

$$v = \frac{C_D}{C_L^{3/2}}\sqrt{\frac{2W}{\rho_0\sigma S}} = \frac{2C_{D_0}}{\left(\dfrac{C_{D_0}}{k}\right)^{3/4}}\sqrt{\frac{2W}{\rho_0\sigma S}}$$

$$= \frac{2 \times 0.023}{\left(\dfrac{0.023}{0.0735}\right)^{3/4}}\sqrt{\frac{2 \times 10{,}000}{0.002377 \times 0.739 \times 255}} = 23.2\ \frac{\text{ft}}{\text{sec}}$$

This completes the basic glide problem: calculation of the glide angle and the sinking speed for shallow paths (i.e., $L = W$ and $\sin\gamma \approx \gamma$). For steeper angles of glide $L = W$, $\cos\gamma$ and the glide angle becomes

$$\tan\gamma = \frac{-1}{C_L/C_D} \tag{3.76}$$

However, this is still not correct; for steep glide angles one must also consider the kinetic energy contribution and use the energy equation Eq. 3.66. The glide angle becomes

$$\tan\gamma = -\frac{C_D/C_L}{1 + \dfrac{V}{g}\dfrac{dV}{dh}} \tag{3.77}$$

In order to obtain the glide angle, the velocity variation with altitude must be specified, which implies that a glide schedule must be also known. Otherwise, the problem becomes a path-integration problem. It is possible, however, to reduce Eq. 3.77 to a point-performance problem if the instantaneous velocity is known. This is achieved, for a constant angle of attack, by differentiating the instantaneous path velocity:

$$V = \sqrt{\frac{2W\cos\gamma}{\rho S C_L}} \tag{3.78}$$

with respect to altitude and determining dV/dh, since the angle of attack is assumed to be constant. One obtains

$$\begin{aligned}\frac{dV}{dh} &= -\frac{1}{2}\frac{V}{\rho}\frac{d\rho}{dh} - \frac{V}{2}\tan\gamma\frac{d\gamma}{dh} = -\frac{1}{2}\frac{V}{\rho}\frac{d\rho}{dh} \\ &- \frac{(V/2)\tan\gamma(d\gamma/dh)}{dh/dt} = -\frac{1}{2}\frac{V}{\rho}\frac{d\rho}{dh}\end{aligned} \tag{3.79}$$

because it was assumed that $\dot{\gamma} = 0$. Thus, the glide angle becomes

$$\tan\gamma = -\frac{C_D/C_L}{1 - \dfrac{V^2}{2g}\dfrac{1}{\rho}\dfrac{d\rho}{dh}} \tag{3.80}$$

Either V or C_L can be eliminated from Eq. 3.80 by means of Eq. 3.78. Since C_L is assumed to be constant, it is seen that the velocity decreases as the altitude decreases (the density ρ increases), and the kinetic energy correction in the denominator becomes increasingly smaller for lower altitudes. Since the density variation is given by different expressions for troposphere and stratosphere, the problem needs to be considered in two steps.

Troposphere In the troposphere, $h < 36{,}089$ ft, the density variation is given by

$$\sigma = (1 - 6.88 \times 10^{-6}h)^{4.256}$$

and the glide angle, Eq. 3.80, becomes

$$\tan \gamma = -\frac{C_D/C_L}{1 + \dfrac{2.928 \times 10^{-5}V^2}{2g\sigma^{.235}}} \tag{3.81}$$

This indicates that in the troposphere, where $1 > \sigma > .297$, the kinetic energy correction in the denominator is small for flight velocities less than 400 to 500 ft/sec.

Stratosphere In the stratosphere, $\sigma < .297$, the density ratio is given by

$$\sigma = .297e^{-4.8\times10^{-5}(h-36089)}$$

which leads to the following expression for the glide angle:

$$\tan \gamma = -\frac{C_D/C_L}{1 + \dfrac{4.8 \times 10^{-5}V^2}{2g}} \tag{3.82}$$

In the stratosphere, where the glide velocities must be high to sustain lift, the kinetic energy contribution is seen to be significant (see Example 3.6).

3.4.2 Glide Range and Endurance

The range and endurance can be calculated with some generality for a glide at constant angle of attack. The shallow glide path and the case without kinetic energy term will appear as a special case. The following set of differential equations are obtained for the range and endurance. For range, the energy equation Eq. 3.2 can be written as

$$\frac{dh}{dt} = -\frac{DV}{W} - \frac{dV^2}{2gdt} \tag{3.83}$$

and if V is eliminated from the $(DV)/W$ term by means of Eqs. 2.10 and 2.12, $x = R$ (range), $\dot{\gamma} \approx 0$ becomes

$$\frac{dR}{L/D} = -dh - \frac{dV^2}{2g} \tag{3.84}$$

The equation for time is obtained from Eq. 2.11, $(T = 0)$

$$\frac{1}{g}\frac{dV}{dt} = -\frac{D}{W} - \sin\gamma \tag{3.85}$$

Eliminating $\sin\gamma = (1/V)(dh/dt)$ (Eq. 2.9), and multiplying by dt, one obtains $(\cos\gamma \approx 1)$

$$\frac{dt}{L/D} = -\frac{dh}{V} - \frac{dV}{g} \tag{3.86}$$

Since $L/D = C_L/C_D$ and $C_D = C_D(C_L)$, it is possible to integrate Eq. 3.84 and Eq. 3.86 if one assumes that the angle of attack is constant (C_L = constant). With the following boundary conditions between any two points

$$R_1 = 0,\ t_1 = 0$$

one gets

$$R = \frac{C_L}{C_D}\left(h_1 - h_2 + \frac{V_1^2 - V_2^2}{2g}\right) \tag{3.87}$$

To integrate Eq. 3.86 for endurance, the velocity will be eliminated by Eq. 3.78, and one obtains for time:

$$t = -\frac{C_L}{C_D}\sqrt{\frac{C_L\rho_0 S}{2W}}\int_1^2 \sqrt{\sigma}dh + \frac{C_L}{C_D}\left(\frac{V_1^2 - V_2^2}{2g}\right) \tag{3.88}$$

As in the previous section, the integral containing the density ratio must be evaluated for two cases:

Troposphere

$$t_{\text{trop}} = \frac{46467}{v_{c_o}}(\sigma_2^{.74} - \sigma_1^{.74}) + \frac{C_L}{C_D}\left(\frac{V_1 - V_2}{g}\right) \tag{3.89}$$

Stratosphere

$$t_{\text{strat}} = \frac{41667}{v_{c_o}}(\sqrt{\sigma_2} - \sqrt{\sigma_1}) + \frac{C_L}{C_D}\left(\frac{V_1 - V_2}{g}\right) \tag{3.90}$$

The following results can be summarized:

1. The range equation is essentially an energy expression. The range is determined by the difference between the initial and final energy heights $h + V^2/2g$. The energy is converted into range by the lift/drag ratio. Thus, the maximum range is obtained, for given initial and final conditions, at an angle of attack that yields E_m.
2. The velocity for maximum range is given by Eq. 3.26:

$$V_{(L/D)_{\max}} = V_{D_{\min}}$$

3. The maximum endurance is essentially determined by $C_D/C_L^{3/2}$, which implies that the path flight velocity is $V_{P_{\min}}$ (see Eq. 3.56). Due to the velocity term, where C_L/C_D appears, the angle of attack for maximum endurance is such that the optimum flight velocity is somewhere between $V_{(L/D)_{\max}}$ and $V_{P_{\min}}$.

EXAMPLE 3.6

The glide range for Example 3.5 will now be reexamined using Eq. 3.87:

$$R = \frac{C_L}{C_D}\left(h_1 - h_2 + \frac{V_1^2 - V_2^2}{2g}\right)$$

The potential energy contribution was calculated in Example 3.5, and it resulted in a range of 23 miles. The kinetic energy contribution is obtained from

$$V_1 = V_{D_{min\,10,000}} = 283\ \frac{\text{ft}}{\text{sec}}$$

$$V_2 = V_{D_{min_0}} = 243\ \frac{\text{ft}}{\text{sec}}$$

and the range increase is given by

$$\frac{C_L}{C_D}\left(\frac{V_1^2 - V_2^2}{2g}\right) = 12.16 \times \frac{283^2 - 243^2}{2 \times 32.2} = 4{,}973\ \text{ft}$$

However, the kinetic energy term becomes significant for a reentry vehicle even with a low $C_L/C_D = 2.5$, which is descending from 80,000 ft at 17,400 ft/sec to a sea level landing at 300 ft/sec. Then, the total glide range is

$$R = \frac{2.5}{5{,}280}\left(80{,}000 + \frac{17{,}400^2 - 300^2}{2 \times 32.2}\right) = 2{,}263\ \text{miles}$$

PROBLEMS

3.1 An aircraft weighing 18,000 lbs, with wing area of 350 ft^2, flies at a true air speed (TAS) of 250 knots. Calculate the aircraft lift coefficient, C_L, and the equivalent air speed (EAS) V_E if the flight altitude is:

a. sea level

b. 20,000 ft

3.2 After the aircraft in Problem 3.1 has consumed 3,000 lbs of fuel, what value of C_L will be required to fly at the same TAS of 250 knots at 20,000 ft altitude? The aircraft in Problem 3.1 is now flying on its landing approach at its maximum lift coefficient $C_{L_{max}}$ of 1.5. If the gross weight is 15,000 lbs, what is the corresponding $V_{E_{min}}$?
Ans: 155 ft/sec.

3.3 It is much safer to fly at a speed marginally above $V_{E_{min}}$ (i.e., $V_{land} = 1.2\ V_{E_{min}}$). Calculate this speed and then calculate the corresponding TAS for landing at an altitude of 6,000 ft.

3.4 The landing speed of an airplane is 10 mph greater than its stalling speed. $C_{L_1} = 1.5$, $C_{L_s} = 1.8$. Find the stalling speed.

3.5 An airplane has the following characteristics:

$$C_{L_{max}} = 1.9$$
$$\alpha_{L=0} = \text{deg } -2$$
$$\text{Lift curve slope a} = 0.075 \text{ per deg}$$
$$V_{s_{max}} = 60 \text{ mph at } 3{,}000 \text{ ft}$$

At what velocity will it fly at sea level if the angle of attack is $\alpha = 1/\text{deg}$?
Ans: 245 ft/sec.

3.6 The aircraft described in Example 3.2 is a T-2B jet trainer equipped with J60-P-6 turbojets, and its empty weight is 10,000 lb. The drag polar given there is for a clean configuration. If the aircraft is loaded with two racks of rockets (rockets + racks = 487 lbs), its parasite area is now 6.25 ft^2 and the span efficiency drops to 0.82. If it carries 2,500 lb fuel, calculate:

a. The new drag polar

b. $(L/D)|_{max}$

c. T_r at minimum drag and at 20,000 ft

3.7 The aircraft in the last problem is equipped with 2 J60-P-6 jet engines and weighs 13,000 lb. The thrust can be approximated by

$$\text{T} = 2600\ \sigma^m$$
$$m = 0.72\ h < 36{,}000 \text{ ft}$$
$$m = 1\ h > 36{,}000 \text{ ft}$$

Find (a) maximum velocity at 25,000 ft (b) its ceiling

3.8 The following data are given for P-3C Orion ASW aircraft:

S = 1300 ft^2
f = 29.1 ft^2
AR = 7.5
W = 128,000 lb
b = 99 ft (wing span)
span efficiency = 0.948 (low speed)

Calculate the thrust horsepower required for 15,000 ft at a speed of 200 knots.
Ans: 5,588 HP.

3.9 An aircraft has the following characteristics:

$$W/S = 40 \text{ lb/ft}^2$$
$$S = 400 \text{ ft}^2$$
$$C_D = 0.018 + 0.062\, C_L^2$$

Calculate:

a. Minimum drag speed, EAS (ft/sec)
b. Minimum power required speed, EAS (ft/sec)
c. Minimum power required
d. Minimum thrust required for steady, level flight

3.10 A transport aircraft cruise weight is about 650,000 lb. Its best cruise speed is 625 mph at 42,000 ft. It is equipped with four JTD9-3 engines. Calculate its drag coefficient and estimate the thrust-specific fuel consumption. Its wing area is 5,500 ft^2.
Ans: $C_D = .0227$.

3.11 Show that

$$V_{P_{\min}} = \left(\frac{4}{3\pi f e}\right)^{1/4} \sqrt{\frac{1}{\sigma}\frac{1}{\rho_0}\frac{W}{b}}$$

where b is the wing span, and

$$P_{\min} = \frac{1.15}{550}\frac{WV_{P_{\min}}}{E_{\mathrm{m}}}$$

3.12 The clean drag polar of a fighter is

$$C_D = 0.04 + 0.09C_L^2$$

Its landing drag polar (flaps, slats and gear down) is

$$C_D = 0.2 + 0.32C_L^2$$

It is equipped with one engine whose thrust can be approximated by

$$T = 15{,}000\ \sigma^m \qquad m = 1\ h > 36{,}000 \text{ ft}$$
$$m = 0.8\ h < 36{,}000 \text{ ft}$$

Its maximum lift coefficient is 2.16. Its take-off speed is the same as its landing speed. Assume that the landing speed is 1.2 V_s. Calculate:

a. T/W at ceiling

b. T/W at takeoff

Ans: a: 0.12; b: 0.613

3.13 Show that for level flight, with the engine thrust inclined at angle α_T to the flight path, the power-on stall velocity is given by

$$V_s = \sqrt{\frac{2W}{\rho S C_{L_{max}}}\left(1 - \frac{T}{W}\sin\alpha_T\right)}$$

Hint: Consider Eqs. 2.9 to 2.13. Derive the full equation.

3.14 A wing is gliding from a height h into a horizontal head wind of V_w mph. If the wing loading is l_w (psf), find an equation for horizontal distance d (ft) that the wing travels before striking ground. Show that

$$d = \frac{C_L}{C_D} h \left[1 - V_w \sqrt{\frac{\rho}{2}\frac{C_L}{l_w}}\right]$$

Assume small angles.

3.15 A jet aircraft is at 20,000 ft with the following characteristics:

$$W = 13{,}000 \text{ lb}$$
$$T = 5{,}200 \text{ lb}$$
$$S = 255 \text{ ft}^2$$
$$AR = 5.07$$
$$C_D = 0.0245 + 0.0765\, C_L^2$$

There is a flame-out and the engine cannot be started. Calculate:

a. Potential maximum glide range

b. Path angle

c. Maximum range sinking speed

d. Horizontal velocity

e. The glide angle at the minimum sinking speed

f. Minimum sinking speed

g. Range at minimum sinking speed

3.16 Calculate the minimum sinking speed and resulting glide range from the data given in Example 3.5.
Ans: $v_{cm} = 20.4$ ft/sec; $R = 19.9$ mi.

3.17 A lifting reentry body with $C_L/C_D = 2.5$ is gliding at 80,000 ft when its flight speed is 17,400 ft/sec. Calculate its flight path angle without and with the kinetic energy term.
Ans: -21.8, $-.1$ deg, resp.

3.18 The drag polar for a glider is given by

$$C_D = 0.017 + 0.021C_L^2$$
$$AR = 20.4$$
$$b = 102 \text{ ft}$$
$$W = 840 \text{ lb}$$
$$h = 35{,}000 \text{ ft}$$

Calculate its range and time to glide to 5,000 ft at constant angle of attack.

3.19 A twin-engine, propeller-driven airplane has the following characteristics:

$$E_m = 12$$
$$f = 17.45 \text{ ft}^2$$
$$S = 500 \text{ ft}^2$$
$$W = 67{,}000 \text{ lb}$$

The pilot is told to hold at minimum fuel flow speed at 5,000 ft. Calculate the holding speed.
Ans: $V_{P_{min}} = 300$ ft/sec.

3.20 A 15,000 lb airplane is coming in for a landing on instrument approach. Its landing drag polar is $C_D = 0.042 + 0.06C_L^2$ and $C_{L\max}$ is 2.9. Calculate the thrust required at the landing speed.
Ans: $T = 1{,}913$ lb.

3.21 An Airbus A330-200, with a crew of 13 and 293 passengers, runs out of fuel (a bad leak) at an altitude of 39,000 ft. Empty weight of the aircraft is 266,000 lb. Determine:

a. Whether this aircraft can glide to the nearest airport, 85 nmi away

b. The flight speed for the maximum range, at altitude, at sea level (Normal landing speed is 250 km/hr, flight manual recommends a flight speed of 170kt for a deadstick landing.)

c. The sinking speed corresponding to *b.* at sea level

d. How much range would be available if the energy term would be included in the calculations.

Ans: a. 129 mi.

3.22 A propeller-driven airplane weighs 4,000 lb, has a wing area of 200 ft with an aspect ratio of 6.5, an efficiency factor of .7, and a zero-lift drag coefficient of .0277. It has two engines, each producing a maximum of 200 HP at propeller efficiency of .82. Estimate the maximum sea-level speed attainable by this aircraft.

3.23 Calculate minimum required power for the aircraft in problem 3.22. *Hint:* Problem 3.11.

Ans: 91.7 HP.

3.24 Verify Eq. 3.81.

3.25 The following is known about a multiengined aircraft:

$$S = 1{,}300 \text{ ft}^2$$
$$f = 29.1 \text{ ft}^2$$
$$AR = 7.5$$
$$\text{span} = 99 \text{ ft}$$
$$\text{span efficiency} = .948$$
$$C_{L_{max}} = 2.6$$
$$W = 128{,}000 \text{ lb}$$

Determine:

a. How much horsepower is required to fly this aircraft at sea level at 200 knots

b. The landing speed

c. The maximum weight to land at 100 knots

3.26 An aircraft has the following characteristics:

$$C_D = .025 + .057 C_L^2$$
$$W = 30{,}000 \text{ lb}$$
$$S = 550 \text{ ft}^2$$
$$AR = 5$$
$$e = .895$$
$$\text{span} = 52.44 \text{ ft}$$

The thrust at 10,000 ft altitude is constant at 8,000 lb. Determine V_{max} at that altitude.
Ans: 805 ft/sec.

3.27 At sea-level testing, a twin-engine jet aircraft was observed to have a maximum velocity of 746 ft/sec. Its weight was 14,000 lb with a wing area of 232 ft^2. Engine data for a JT15 engine indicated that the thrust would be, at that altitude and speed, about 1,575 lb/engine. Estimate the parasite drag coefficient.
Ans: .0205

3.28 A twin-engine jet utility aircraft, having been in the holding pattern (at minimum drag speed to conserve fuel) at 5,000 ft, was instructed to reduce velocity by 90 ft/sec prior to commencing landing approach. Aircraft data:

$$W = 10{,}254 \text{ lb}$$
$$S = 260 \text{ ft}^2$$
$$AR = 5.4$$
$$e = .87$$
$$k = .0678$$
$$C_{L_{max,flaps}} = 2.7,\ C_{L_{max,clean}} = 1.7$$
$$C_{D_0} = .017$$

Determine:

a. The amount of thrust change (added or reduced)

b. Whether the resulting flight speed is a safe flying speed

3.29 An aircraft has the following characteristics:

$$C_D = .025 + .057C_L^2$$
$$W = 30{,}000 \text{ lb},\ S = 550 \text{ ft}^2$$
$$AR = 5,\ e = .895$$
$$\text{span} = 52.44 \text{ ft},\ V = 320 \text{ ft/sec}$$

Thrust and chord are along fuselage centerline, $\alpha_{0L} = 0$, wing quarter chord sweep is 12°, twin tail.

a. Obtain the equations of motion when flying up in a perfectly vertical direction (*Hint:* use Eqs. 2.9 to 2.12).

b. What is the thrust required at sea level when this aircraft is in steady vertical flight (i.e., apply the equations obtained in part a)?

Ans: 31,670 lb

3.30 An A7E has flame-out at 15,000 ft and cannot be started. It has the following characteristics:

$$C_D = .015 + .1055C_L^2$$
$$C_{L_{mclean}} = 1.6$$
$$C_{L_{mfl,slats}} = 2.5$$
$$W = 25{,}000 \text{ lb}, \ S = 375 \text{ ft}^2$$

Determine to sea level:

a. Range at minimum sink rate

b. Rate of sink at touchdown

c. Touchdown velocity

d. Estimate time to fly

3.31 It is considered that a single-engine jet aircraft, with the following data, be uprated by adding another identical engine. Assuming that the second engine does not affect the aerodynamics, how much can the aircraft weight be increased to reach the same ceiling? The aircraft has the following parameters: equivalent parasite area 4.84 ft², full span flaps, $AR = 8$, $W = 11{,}000$ lb, $S = 242$ ft², $T = 1{,}000$ lb, $e = .663$, concrete runway, hot standard day.

3.32 Using Eq. B.2 in Appendix B, show that the minimum drag coefficient $C_{D_{min}}$ is now given by

$$C_{D_{min}} = (k + k_1)C_L^2 - kC_{L_o}$$

and $L/D|_{max}$ becomes

$$\left.\frac{L}{D}\right|_{max} = \frac{1}{2\sqrt{(C_{D_0} + kC_{L_o}^2)(k + k_1)} - 2kC_{L_o}^2}$$

4

Climbing Flight

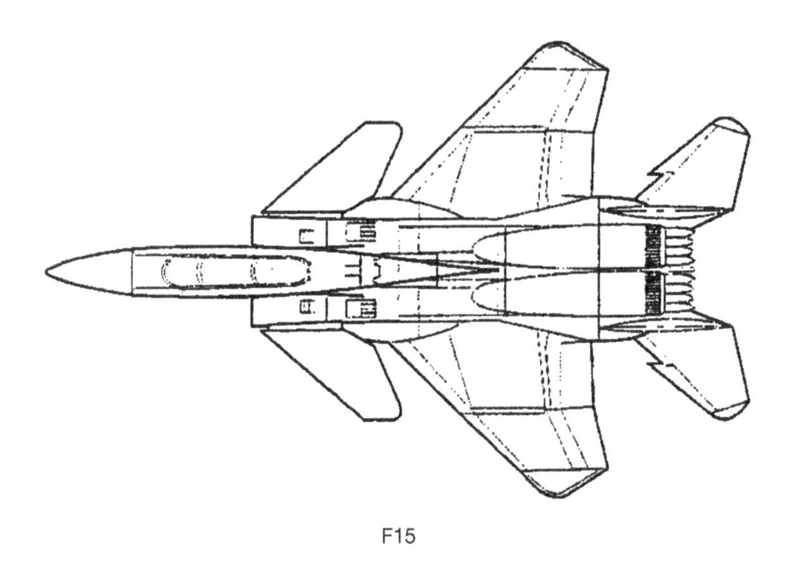

F15

4.1 GENERAL

Climbing performance of an aircraft can be viewed as an energy exchange between the powerplant and the kinetic and potential energies of the aircraft. If the aircraft is in steady, level flight and the power is increased to exceed the amount needed for sustained level flight, then more work is being done on the aircraft than is required to overcome the drag. As a result, the kinetic or potential energy, or both, must increase. Which one increases depends on how the aircraft is being operated.

Keeping the angle of attack constant will tend to increase V until a new $T = D$ balance is achieved. The new increased V, at constant α (and C_L), will produce excess lift, which in turn leads to an increase in potential energy. Moving the control stick to decrease the angle of attack at constant power will decrease C_L and C_D, which at constant power will lead to an increase in V. Since the total lift will remain approximately constant, most of the energy change will go into an increase of the kinetic energy. If the velocity is kept at some constant value, then the excess power will be converted into an increase in potential energy and the aircraft will climb.

The fundamental performance equation, Eq. 3.4,

$$\frac{dh}{dt} = \frac{P_a - DV}{W} \frac{1}{\left[1 + \frac{V}{g}\frac{dV}{dh}\right]}$$

is again appropriate as the basic energy equation for the aircraft. Assuming that $P_a > DV$ (excess of power), dh/dt is positive and is taken as rate of climb (usually given in feet per minute, fpm). If $P_a < DV$, then dh/dt is negative, and the problem is an extension of the gliding flight with partial power on. Normally it is assumed that climb takes place at essentially constant velocity, which justifies then omitting the kinetic energy correction factor $1 + (V/g)(dV/dh)$ in the denominator. This works well for most propeller-driven aircraft or aircraft with modest thrust-to-weight ratio. For high-performance aircraft ($T/W = O(1)$), this assumption needs to be reevaluated. For example, at high speeds and shallow climb angles, dV/dh may be small but the entire term may still be significant.

4.2 RATE OF CLIMB, CLIMB ANGLE

For the time being, it will be assumed that the climb will take place at constant velocity, or at such a rate that the term dV/dh is small. Thus, one obtains immediately the following equations.

The *rate of climb* is given by the following equations (see also Figure 4.1):

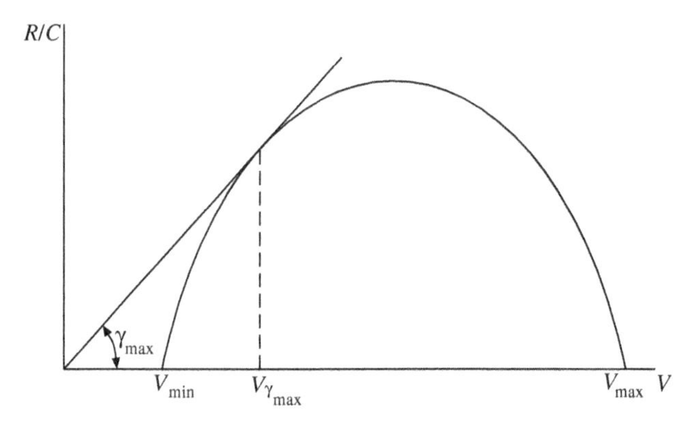

Figure 4.1 Aircraft Climb Performance

$$\frac{dh}{dt} \equiv v = \left.\frac{P_a - DV}{W}\right|_{\text{prop}} \tag{4.1}$$

$$\frac{dh}{dt} \equiv v = \left.\frac{(T - D)V}{W}\right|_{\text{jet}} \tag{4.2}$$

and the *climb angle* is established by means of Eq. 2.9, which gives

$$\frac{v}{V} = \sin\gamma = \left.\frac{P_a - DV}{VW}\right|_{\text{prop}} \tag{4.3}$$

$$\frac{v}{V} = \sin\gamma = \left.\frac{T - D}{W}\right|_{\text{jet}} \tag{4.4}$$

Since the rate of climb is usually given in units of ft/min (fpm), Table 4.1 will give the appropriate conversion factors: dh/dt in fpm, T, D, W in lb.

TABLE 4.1 Climbing Conversion Factors

dh/dt	P	V
$\left(\frac{P_a - DV}{W}\right)$ 33000	HP	ft/sec
$\left(\frac{T - D}{W}\right)$ $60V$	HP	ft/sec
$\left(\frac{T - D}{W}\right)$ $88V$	HP	mph
$\left(\frac{T - D}{W}\right)$ $101.4V$	HP	knots

For any instantaneous values of power, drag, and velocity, Eqs. 4.1 to 4.4 give the aircraft rate of climb, v, and the local inclination of the flight path to the horizon, sin γ (see Figure 2.1). For practical problems, similarly to calculating the maximum velocities in Chapter 3, climb calculations can also be divided into exact and approximate approaches. The exact method depends on having information available in the form of Figures 3.3 or 3.7 and calculations proceed straightforward according to Eqs. 4.1 to 4.4. This will be the focus of the current section. Approximate methods will be taken up later.

Now further subdividing:

A. *Jets* For the climb angle, $T - D$ can be calculated or read directly for a range of values of V from Figures 3.2 or 3.4 and then divided by the weight W. The maximum value and the corresponding flight velocity V are found from the resulting table of values or from a graph of sin γ vs V (see Example 4.1). If thrust is independent of velocity, then the flight velocity for the steepest climb angle occurs at $V_{D_{min}}$ and E_m. The rate of climb is obtained from the same table by multiplying the values of $(T - D)$ by corresponding flight velocity V.
B. *Propeller aircraft* For rate of climb, $(P_a - P_r)$ can be read directly for a range of values of V from Figure 3.7 and then divided by weight W. The maximum value and corresponding flight velocity V are found again from the resulting table or a graph of dh/dt vs V. The climb angle is obtained from the same table by dividing the rate of climb values by corresponding flight velocity V. See also Example 4.1 and Table 4.2 for more details.

This approach to either type of aircraft is straightforward and simple but requires that Figures 3.3 or 3.7 or equivalent data $(T - D)$ in tabular form be already available.

Figure 4.1 shows, for constant altitude, essential features of the climb performance. In addition, the maximum rate of climb can be plotted as a function of altitude, as shown in Figure 4.2, providing another way to determine aircraft ceiling.

The absolute ceiling is defined by the altitude where the rate of climb approaches zero and it occurs in the stratosphere (above 36,000 ft) for jet-powered aircraft, and usually in the tropopause (below 36,000 ft) for propeller-driven aircraft. From an operational point of view, another ceiling is defined: the service ceiling. This occurs when the maximum rate of climb is 100 fpm. Although the ceiling of the propeller-driven aircraft tends to occur below tropopause, its flight path usually exhibits

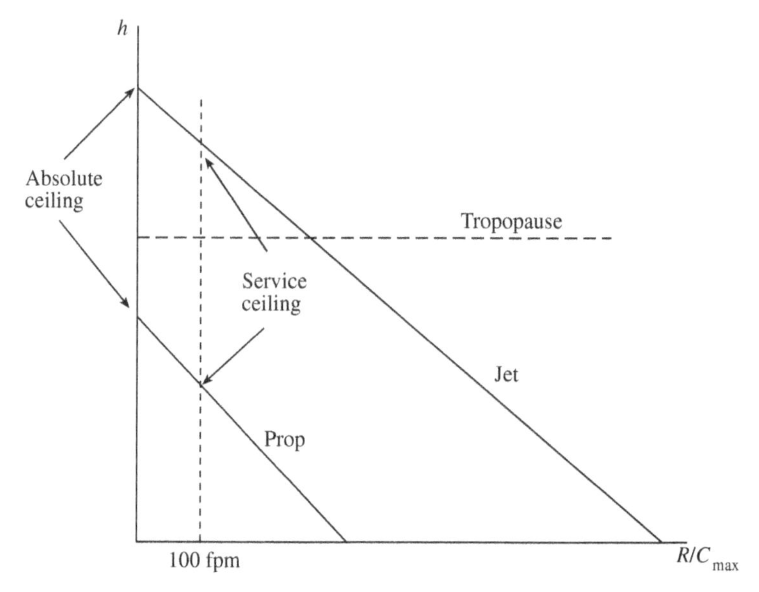

Figure 4.2 Maximum Climb Rate and Ceiling

a steeper climb angle up to almost the service ceiling. Some notable exceptions are found for very-high-performance aircraft with a thrust weight ratio of near unity, or higher. In that case, extremely high climb angles approaching 90 degrees can be obtained.

NOTE

The calculations described on immediately preceeding pages are simple and straightforward but are entirely dependent on the type and quality of drag and thrust data. So are the results. Chapter 3 with all the appendix information is a reminder that much work is required to prepare accurate drag (i.e., C_{D0}) and thrust data for predicting reliable performance information. Chapter 3 and the appendices will provide sufficient aerodynamics and propulsion information only for basic simple cases. Rather, the *basics* are reviewed in Chapter 3, which should be of help in deciding the necessary course of action and provide a warning that all material beyond Chapter 3 is dependent on Chapter 3.

4.3 TIME TO CLIMB

An important aspect of the aircraft climb problem is the amount of time needed to climb from one given altitude h_1 to another altitude h_2. The aircraft rate of climb, or the vertical velocity is defined by

$$v \equiv \frac{dh}{dt} \equiv R/C$$

and the time to climb can be determined from

$$t_2 - t_1 = \int_{h_1}^{h_2} \frac{dh}{v} \tag{4.5}$$

The burden of the problem lies in finding a suitable expression for the rate of climb and integrating this between any two desired altitudes. The last section and the developments in Section 3.3.4 have indicated that the rate of climb for a propeller-driven aircraft must be obtained by numerical means, implying also that the time to climb must be evaluated by numerical integration techniques. As will be seen, for jet aircraft, some analytical expressions are available but the resulting integral expression in Eq. 4.5 turns out to be unduly cumbersome and is best evaluated by numerical techniques.

There are two commonly used methods for calculating the time to climb to altitude. The more precise method for any type of aircraft consists of evaluating the rate of climb v for several altitudes. The area under the $1/v$ curve then gives the required time for climbing between any two altitudes. See Figure 4.3 and Example 4.1. However, as developed by approximate methods in Chapter 3, some simplifications are possible for rate of climb of the jet aircraft (see also the next section).

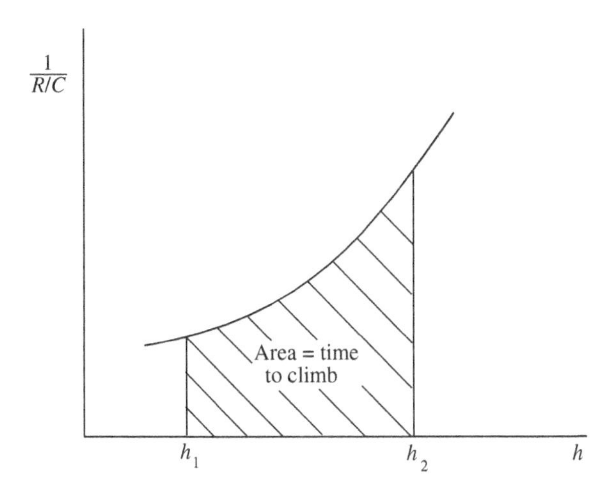

Figure 4.3 Calculation of Time to Climb

The other method is an approximate one since it assumes the rate of climb versus altitude curve is (at least piecewise) linear:

$$v - v_1 = \frac{v_2 - v_1}{h_2 - h_1}(h - h_1) \tag{4.6}$$

When Eq. 4.6 is substituted into Eq. 4.5, the following elementary integral is obtained, which yields

$$t = \int_{h_1}^{h_2} \frac{(h_2 - h_1)\, dh}{v_1 h_2 - v_2 h_1 + h\,(v_2 - v_1)} = \frac{h_2 - h_1}{v_2 - v_1} \ln \frac{v_2}{v_1} \tag{4.7}$$

where it has been assumed, without any loss of generality, that $t_1 = 0$. Eq. 4.7 can be further simplified if the following expansion is used for the logarithm:

$$\ln x = 2\left[\frac{x - 1}{x - 2} + \frac{1}{3}\left(\frac{x - 1}{x + 2}\right)^3 + \ldots\right], x > 0 \tag{4.8}$$

which gives for the time

$$t = \frac{2(h_2 - h_1)}{v_1 + v_2} \tag{4.9}$$

Either of Eqs. 4.7 or 4.9 can be applied in a stepwise manner to calculate the rate of climb over linear portions of the $v - h$ curve if the altitude range is subdivided into suitable segments over which v is linear with altitude. The sum of the times over the individual segments is then the total time to climb.

Since the aircraft can consume considerable quantities of fuel during the climb to high altitudes, the effect of the weight change can be accounted for in an approximate manner in the stepwise calculation if the fuel consumption rate is multiplied by the calculated climb time and the total weight is then reduced accordingly for the following altitude increment. Moreover, such a stepwise calculation also permits allowance for thrust reduction due to the altitude (e.g., Eq. 3.33), which further improves the approximations made above and also yields an estimate for the fuel used in the climb.

The basic methodology for climb problems just discussed is described in Example 4.1. Since the calculations are extensive, the details are shown only for the sea-level altitude. The results are shown in Figure 4.5 for a full spectrum of altitude and velocity. Although the example has been carried out for the propeller-driven aircraft, the method is fully valid for a jet aircraft.

EXAMPLE 4.1

For the propeller-driven aircraft discussed in Example 3.4, determine the climb performance and ceiling for the weight $W = 9{,}117$ lb.

$$C_D = .024 + .0535C_L^2$$

$$AR = 7$$

$$e = .85$$

The rate of climb is determined by use of Eq. 4.1:

$$v \equiv R/C = \frac{P_a - P_r}{W}$$

and the climb angle from

$$\sin \gamma = \frac{R/C}{V}$$

To find the desired performance over the pertinent range of velocities and to establish the maximum values, the calculations are best carried out in a tabular format as shown in Table 4.2. The P_a and P_r values are obtained from Example 3.4 with the suitable range of velocities taken from Figure 3.7. Such a table of values needs to be calculated for a number of altitudes in order to estimate the aircraft ceiling. Table 4.2 shows the values calculated at sea level conditions.

Figure 4.4 shows the rate of climb plotted against the aircraft flight velocity. The maximum rate of climb can be read from the top of the graph to be 17.5 ft/sec (1050 fpm) or estimated from the table to be about 1040 fpm at 220 ft/sec flight speed. A tangent drawn from the origin to the rate of climb curve gives for the maximum path angle 5.3° at about 170 ft/sec.

TABLE 4.2 Rate of Climb Calculation Procedure

V mph	V ft/sec	P_a HP	P_r HP	Excess P $P_a - P_r$	R/C fpm	γ deg
50	73	217	315	−98		
75	110	330	227	103	372	3.2
100	147	413	204	208	753	4.9
125	183	495	220	275	995	5.2
150	220	555	269	286	1035	4.5
175	256	593	352	240	869	3.2
200	293	626	471	155	561	1.8
225	330	648	629	19	69	0.2
350	367	661	831	−170		

Although Figure 4.4 provides a quick overview of the climb performance at one altitude, it is limited and impractical because it is difficult to read the values with sufficient precision from the figure. In any case, the information is already contained in Table 4.2 in adequate detail.

A much more useful representation is found in Figure 4.5, which shows the excess power as a function of velocity for several selected altitudes. Since the flight path angle is usually of no great interest, and since only the maximum rate of climb is of general value, the excess power plot saves extra calculations. Moreover, where the excess power vanishes, one finds either $V_{\max}$ or $V_{\min}$ for that particular altitude.

The maximum rate of climb values for several altitudes are shown plotted in Figure 4.5. The altitude where the maximum rate of climb approaches zero is the aircraft absolute ceiling, in this case, at about 19,500 ft altitude. The curve of maximum ratio of climb is usually determined for several altitudes below the ceiling and then is extrapolated to $R/C = 0$. The extrapolation is assisted by plotting the $V_{\max}$ and $V_{\min}$ curves and also by determining the locus of $R/C_{\max}$ from the excess power curves. Thus, Figure 4.5, which is based on the same calculation as Figure 4.4 but conveys more information in a more precise manner, is to be preferred for practical calculations.

NOTE

Multiplying the excess power curves in Figure 4.5 by V/W yields curves for the specific excess power P_s. It is easily seen that, at a given altitude, the maximum rate of climb is determined by either the maximum value of the excess power curve, which is

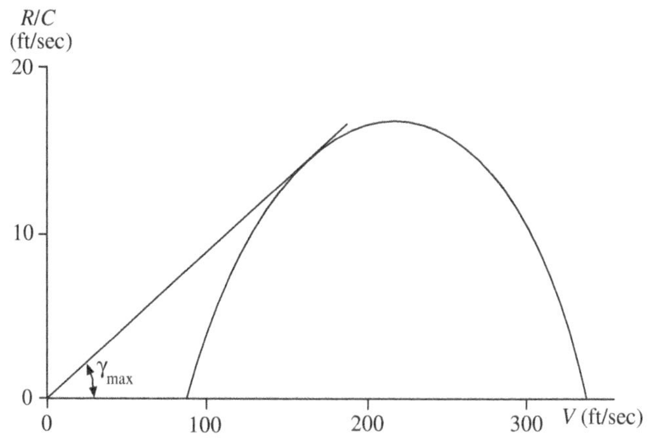

Figure 4.4 Rate of Climb

$$\frac{\partial}{\partial V}[P_a - P_r]_{h=c} = 0$$

or

$$\frac{\partial}{\partial V}\left[\frac{(P_a - P_r)V}{W}\right]_{h=c} = 0$$

This fully anticipates the minimum-time-to-climb process discussed in Chapter 8 under "Energy Methods" (see Eqs. 8.7 and 8.8).

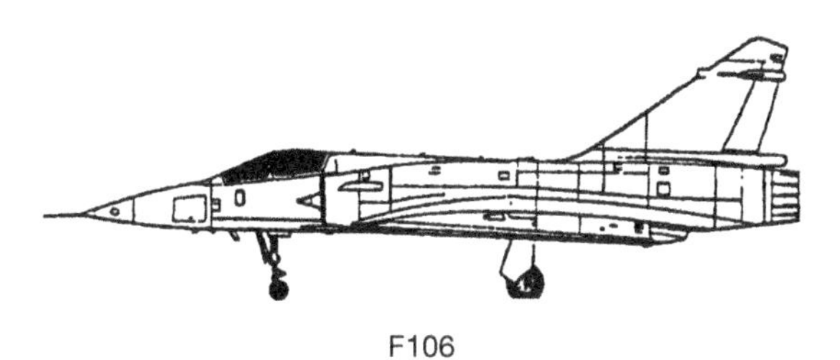

F106

As the last item in this example, the time to climb from sea level to 15,000 ft will be determined. Assume that the weight remains constant at 9,117 lb.

The time to climb can be calculated, with best accuracy, from Eq. 4.5, by first plotting $1/v_c = 1/(R/C)$ from Figure 4.5 or Table 4.2 the original data, as shown in Figure 4.3. Numerical evaluation of the area under the $1/v_c$ curve gives

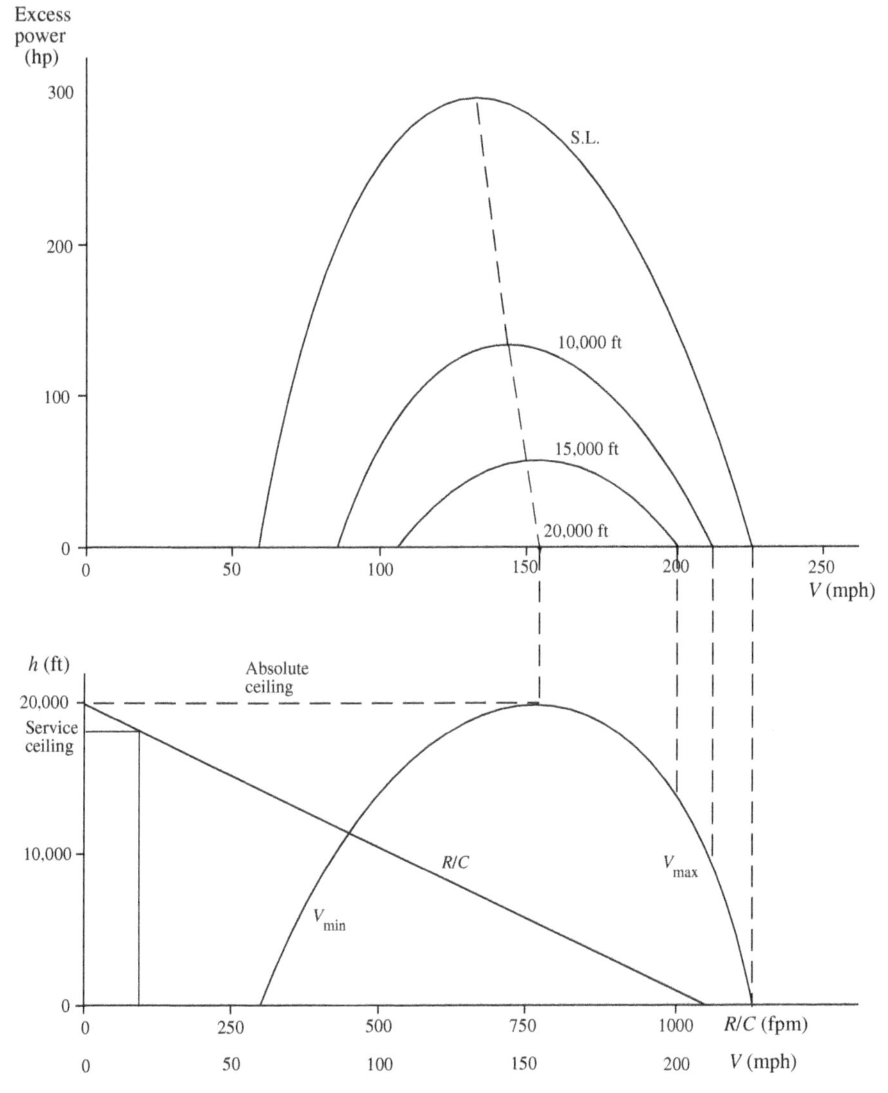

Figure 4.5 Climb Performance

$$t_{15} = \int_0^{15} \frac{dh}{v_c} = 1{,}802 \text{ sec}$$

For comparison, the time is also evaluated from Eqs. 4.7 and 4.9, respectively:

$$t = \frac{h_2 - h_1}{v_2 - v_1} \ln \frac{v_2}{v_1} = \frac{15{,}000}{17.5 - 3.7} \times \ln \frac{17.5}{3.7} = 1{,}689 \text{ sec}$$

and

$$t = \frac{2(h_2 - h_1)}{v_1 + v_2} = \frac{2 \times 15{,}000}{3.7 + 17.5} = 1{,}415 \text{ sec}$$

It is seen that Eq. 4.7 gives a fair approximation, but Eq. 4.9 would miss the answer by about 27 percent. It should be pointed out that assuming the aircraft weight to be constant introduces an error in the answers that can be removed by evaluating, in stepwise manner, the fuel flow during the climb and recalculating the values in Table 4.2. It should be noticed also that there is an implicit assumption made in this procedure that $1/v_c$ is evaluated from a sequence of $R/C|_{max}$ values. This, in turn, leads to a climb with velocity increasing toward ceiling (see Figure 4.5). If climb time at constant velocity is desired, the consistent values of R/C must be chosen from Figure 4.5. All these will not be $R/C|_{max}$.

4.4 OTHER METHODS

In case there is a need for a fast and closed form of solution, the approximate methods described in this section are useful. It should be recognized that the approximate methods work well for jet aircraft but are not always suitable for propeller-driven aircraft, as explained in Chapter 3.

4.4.1 Shallow Flight Paths

For a jet aircraft, the climb angle can be developed from Eq. 4.4 and the nondimensional form of the drag, Eq. 3.36, as follows (with $n = 1$):

$$\sin \gamma = \frac{T}{W} - \frac{D}{W} = \frac{T}{W} - \frac{1}{2E_m}\left(\overline{V}^2 + \frac{1}{\overline{V}^2}\right) \tag{4.10}$$

where again

$$\overline{V} = \frac{V}{V_{D_{\min}}} = \frac{V}{\sqrt{\dfrac{2W}{\rho S}\left(\dfrac{k}{C_{D_0}}\right)^{1/4}}}$$

Multiplying the above by V, one obtains the rate of climb, or vertical velocity

$$v = \frac{dh}{dt} = V \sin \gamma \equiv R/C \tag{4.11}$$

which becomes, in nondimensional form, after dividing by $V_{D_{\min}}$

$$\bar{v} = \frac{v}{V_{D_{\min}}} = \frac{T\bar{V}}{W} - \frac{1}{2E_m}\left(\bar{V}^3 + \frac{1}{\bar{V}}\right) \tag{4.12}$$

The maximum values may be obtained by means of simple differentiation, but it must be kept in mind that the expression for the thrust may be a function of velocity. As an example for this procedure using the thrust expression given by Eq. 3.33, which is independent of velocity, one gets

$$\sin \gamma = \frac{T_o\sigma^m}{W} - \frac{1}{2E_m}\left(\bar{V}^2 + \frac{1}{\bar{V}^2}\right) \tag{4.13}$$

and

$$\bar{v} = \frac{T_o\sigma^m\bar{V}}{W} - \frac{1}{2E_m}\left(\bar{V}^2 + \frac{1}{\bar{V}^2}\right) \tag{4.14}$$

The maximum values can be obtained in a straightforward manner by differentiating these two equations with respect to $\bar{V}$

$$\frac{d \sin \gamma}{d\bar{V}} = -\frac{1}{E_m}\left(\bar{V} + \frac{1}{\bar{V}^3}\right) = 0 \tag{4.15}$$

or $\bar{V} = 1$, which states that the steepest climb occurs at

$$V = V_{D_{\min}} \tag{4.16}$$

or that

$$\sin \gamma_{\max} = \frac{T_o\sigma^m}{W} - \frac{1}{E_m} \tag{4.17}$$

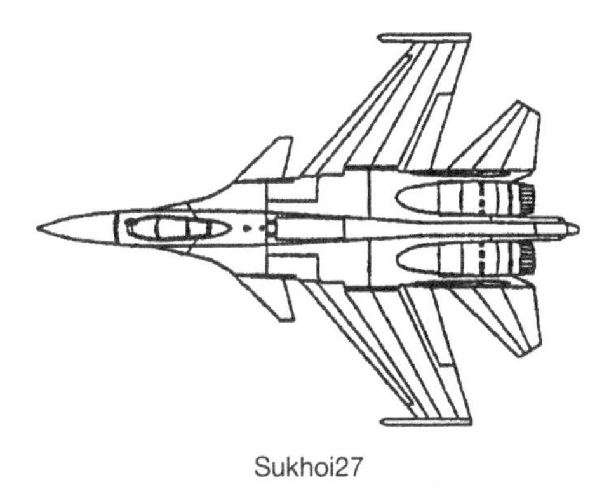

Sukhoi27

EXAMPLE 4.2

For a high-performance aircraft with a thrust-to-weight ratio of 0.9 and a L/D_{max} of 10, calculate the maximum climb angle. A straightforward substitution into Eq. 4.17 with

$$\frac{T_o \sigma^m}{W} = 0.9$$

and

$$E_m = 10$$

gives

$$\sin \gamma_{max} = 0.9 - \frac{1}{10} = 0.8$$

Therefore

$$\gamma_{max} = 53.1°$$

which is a rather steep climb angle and exceeds the concept of a shallow path angle.

The condition for fastest climb is obtained from Eq. 4.12:

$$\frac{d\bar{v}}{d\bar{V}} = \frac{T}{W} - \frac{1}{2E_m}\left(3\bar{V}^2 - \frac{1}{\bar{V}^2}\right) = 0 \tag{4.18}$$

which gives, after some rearrangement,

$$3\bar{V}^4 - 2\bar{T}\bar{V}^2 - 1 = 0 \tag{4.19}$$

with

$$\bar{T} = \frac{TE_m}{W}.$$

This biquadratic equation can be easily solved (see Eqs. 3.45 and 3.46) to give one physically possible solution

$$\bar{V}_{\bar{v}_{\max}} = \frac{\sqrt{\bar{T} + \sqrt{\bar{T}^2 + 3}}}{\sqrt{3}} \tag{4.20}$$

Substituting back into Eq. 4.12 gives, by use of Eq. 4.20, for the maximum rate of climb

$$\begin{aligned} \bar{v}_{\max} &= \frac{1}{2E_m}\left(\bar{T}\bar{V} - \bar{V}^3 + \frac{1}{\bar{V}}\right) \\ &= \frac{0.3849}{E_m}\sqrt{\bar{T} + \sqrt{\bar{T}^2 + 3}}\left(2\bar{T} - \sqrt{\bar{T}^2 + 3}\right) \end{aligned} \tag{4.21}$$

The result given by Eq. 4.16 should have been anticipated, as the climb angle for jet aircraft is determined by the excess power and occurs, for the constant thrust case, at $V_{D_{\min}}$ (see also Example 3.3). If the thrust varies with velocity, the steepest climb location decreases or increases beyond $V_{D_{\min}}$ depending on whether the $T - V$ curve slope is negative or positive, respectively. This is also seen if Eq. 3.34 is used for the thrust in Eqs. 4.10 and 4.12. Only the results are given below; the details are left as an exercise. Upon substituting Eq. 3.34 into 4.10, the climb angle is obtained as

$$\sin\gamma = \frac{(A + BV^2)\sigma^m}{W} - \frac{1}{2E_m}\left(\overline{V}^2 + \frac{1}{\overline{V}^2}\right)$$

$$= \frac{1}{2E_m}\left(\frac{2A\sigma^m E_m}{W} + \frac{2V^2 B\sigma^m}{\frac{2W}{\rho S}\frac{\sqrt{k}}{C_{D_0}}\rho S C_{D_0}} - \overline{V}^2 - \frac{1}{\overline{V}^2}\right)$$

$$+ \frac{1}{2E_m}\left[2\overline{A} - \overline{V}^2(\overline{B} - 1) - \frac{1}{\overline{V}^2}\right] \tag{4.22}$$

where

$$\overline{A} = \frac{A\sigma^m E_m}{W}$$

$$\overline{B} = \frac{2B\sigma^m}{\rho S C_{D_0}}$$

Similarly, the rate of climb is obtained from Eq. 4.12:

$$\overline{v} = \frac{1}{2E_m}\left[2\overline{AV} - \overline{V}^3(\overline{B} - 1) - \frac{1}{\overline{V}}\right] \tag{4.23}$$

The maximum values of the climb angle and climb velocity are

$$\sin\gamma_{\max} = \frac{1}{2E_m}\left[2\overline{A}^2 - (1 - \overline{B})^{3/2} - \frac{1}{\sqrt{1 - \overline{B}}}\right] \tag{4.24}$$

$$\overline{V} = (1 - \overline{B})^{1/4} \tag{4.25}$$

The maximum rate of climb is found to be

$$\overline{v}_{\max} = \frac{1}{2E_m}\left[2\overline{AV} - \overline{V}^3(1 - \overline{B}) - \frac{1}{\overline{V}}\right] = \frac{2\overline{V}}{3E_m}\left[\overline{A} - \frac{1}{\overline{V}^2}\right]$$

$$= \frac{2\sqrt{\overline{A} + \sqrt{\overline{A}^2 + 3(1 - \overline{B})}}}{3\sqrt{3(1 - \overline{B})}E_m}\left[2\overline{A} - \sqrt{\overline{A}^2 + 3(1 - \overline{B})}\right] \tag{4.26}$$

Eqs. 4.22 and 4.26 reduce for the case of $B = 0$ (thrust is independent of the velocity) to Eqs. 4.17 and 4.21, respectively.

EXAMPLE 4.3

An aircraft has the following characteristics:

$$W = 36{,}000 \text{ lb}$$
$$S = 450 \text{ ft}^2$$
$$T_0 = 6{,}000 \text{ lb}$$
$$C_D = 0.014 + 0.05\, C_L^2$$

If its thrust is assumed to be constant and it climbs at a speed of 300 ft/sec at a shallow angle, calculate:

a. Angle of climb
b. Rate of climb
c. Maximum angle of climb
d. Maximum rate of climb

a. The problem is most easily solved if the climb angle is calculated first from Eq. 4.10 by establishing

$$\frac{T}{W} = \frac{6{,}000}{36{,}000} = \frac{1}{6}$$

$$V_{D_{\min}} = \sqrt{\frac{2W}{\rho S}\left(\frac{k}{C_{D_0}}\right)^{1/4}} = \sqrt{\frac{2 \times 36{,}000}{0.002377 \times 450}\left(\frac{0.05}{0.014}\right)^{1/4}}$$
$$= 356.6\ \frac{\text{ft}}{\text{sec}}$$

$$\bar{v} = \frac{V}{V_{D_{\min}}} = \frac{300}{356.6} = 0.841$$

$$E_m = \frac{1}{2\sqrt{kC_{D_0}}} = \frac{1}{2\sqrt{0.014 \times 0.05}} = 18.89$$

Therefore

$$\sin\gamma = \frac{1}{6} - \frac{1}{2 \times 18.89}\left(0.841^2 + \frac{1}{0.841^2}\right) = 0.1105$$

whence

$$\gamma = 6.34°$$

b. The rate of the climb can be calculated from Eq. 4.12, or, from a simpler expression,

$$v = V \sin \gamma = 300 \times 0.1105 = 33.15 \frac{\text{ft}}{\text{sec}} = 1989 \text{ fpm}$$

It should be noted that the climb does not take place at $(L/D)_{\max}$ (it is only a convenient means for establishing Eq. 4.10), but at

$$C_L = \frac{2W}{\rho_o V^2 S} = \frac{72{,}000}{0.002377 \times 300^2 \times 450} = 0.748$$

which gives an L/D of

$$\frac{C_L}{C_D} = \frac{0.748}{0.014 + 0.05 \times 0.748^2} = 17.8$$

The maximum angle of climb is obtained from Eqs. 4.16 and 4.17 as follows:

$$\gamma_{\max} = \sin^{-1}\left(\frac{T}{W} - \frac{1}{E_m}\right) = \sin^{-1}\left(0.1667 - \frac{1}{18.89}\right) = 6.53°$$

and occurs at $V = V_{D_{\min}} = 356.6$ ft/sec and at $(L/D)_{\max}$.
Maximum rate of climb is obtained from Eq. 4.21 with

$$\overline{T} = \frac{T}{W} E_m = \frac{6{,}000}{36{,}000} \times 18.89 = 3.15$$

$$\overline{v}_{\max} = \frac{0.3849}{18.89} \sqrt{3.15 + \sqrt{3.15^2 + 3}} \times (2 \times 3.15 - \sqrt{3.15^2 + 3}) = 0.1432$$

where

$$v_{\max} = 0.1432 V_{D_{\min}} = 0.1432 \times 356.6 = 51.05 \frac{\text{ft}}{\text{sec}} = 3{,}063 \text{ fpm}$$

It is obtained at the following flight speed (Eq. 4.20):

$$\overline{V}_{\bar{v}_{max}} = \frac{\sqrt{\overline{T} + \sqrt{\overline{T}^2 + 3}}}{\sqrt{3}} = \frac{\sqrt{3.15 + \sqrt{3.15^2 + 3}}}{\sqrt{3}} = 1.4994$$

or

$$V = 1.4994 V_{D_{min}} = 1.4994 \times 356.6 = 534.7 \ \frac{\text{ft}}{\text{sec}}$$

and at

$$C_L = \frac{72{,}000}{0.002377 \times 534.7^2 \times 450} = 0.235$$

and corresponding

$$\frac{L}{D} = 14.03$$

This concludes the discussion of shallow flight paths where $n \approx 1$, or $L \approx W$. From a practical engineering point of view, many of the climb problems can be analyzed by use of the techniques presented in this section. In cases where the thrust-weight approaches unity, or the flight path angle is very steep, or both, the above approach fails to take into account the variation of the velocity with the changing altitude. Even this omission will not cause great errors in the flight path angle, but the errors become significant in the calculation of the optimum velocity.

In the next section, a technique will be discussed for evaluating performance at steep angles where the load factor $n \neq 1$.

4.4.2 Load Factor $n \neq 1$*

For small climb angles it has been assumed that $n = L/W = 1$, or in other words, $\cos \gamma = 1$. At steeper climb angles, this approximation ceases to be accurate. An improved solution is obtained if one considers Eq. 2.12 with $\epsilon = 0$ and $\dot{\gamma} = 0$. This gives

$$L = W \cos \gamma \tag{4.27}$$

$$n - \cos \gamma = 0 \tag{4.28}$$

Eq. 2.11 supplies the equation tangent to the flight path, with $\dot{V} = 0$, which can be immediately restated in nondimensional form, similar to Eq. 4.10, as follows

$$\sin \gamma = \frac{T}{W} - \frac{1}{2E_m}\left(\overline{V}^2 + \frac{n^2}{\overline{V}^2}\right) \tag{4.29}$$

Eliminating n^2, and using $\cos^2 \gamma + \sin^2 \gamma = 1$, one obtains

$$\overline{V}^4 - 2\overline{V}^2 E_m \left(\frac{T}{W} - \sin \gamma\right) + 1 - \sin^2 \gamma = 0 \tag{4.30}$$

This equation can be solved for either $\overline{V}$ or $\sin \gamma$. In the first case, the solution depends on the type of the thrust expression available, i.e. is it a function of velocity (Eq. 3.34). The case for $\sin \gamma$ can be solved without recourse to an explicit thrust statement.

To illustrate the method, it will be assumed that the thrust is independent of the velocity and Eq. 3.33 holds. Use of Eq. 3.34 will be left as an exercise. Solving for the velocity, one gets

$$\overline{V} = E_m \left(\frac{T}{W} - \sin \gamma\right) \pm \sqrt{E_m^2 \left(\frac{T}{W} - \sin \gamma\right)^2 - (1 - \sin^2 \gamma)} \tag{4.31}$$

Eq. 4.31 gives, for a fixed thrust-weight ratio and flight path angle, the required flight path velocity. Both signs give physically possible results. The positive sign gives the higher-speed solution (see Section 3.3.3). To obtain γ and v, Eq. 4.30 can be rewritten as follows:

$$\sin^2 \gamma - 2\sin \gamma E_m \overline{V}^2 + 2E_m \frac{T}{W} \overline{V}^2 - 1 - \overline{V}^4 = 0 \tag{4.32}$$

Solving for $\sin \gamma$ gives

$$\sin \gamma = \overline{V}^2 E_m \pm \sqrt{E_m^2 \overline{V}^4 - (2\overline{T}\overline{V}^2 - 1 - \overline{V}^4)} \tag{4.33}$$

where

$$\bar{T} = \frac{T}{W} E_m \tag{4.34}$$

(for any expression of thrust), from which the rate of climb is obtained as

$$\bar{v} = \frac{v}{V_{D_{\min}}} = \bar{V} \sin \gamma = \bar{V}^3 E_m \pm \bar{V}\sqrt{E_m \bar{V}^4 - (2\bar{T}\bar{V}^2 - 1 - \bar{V}^4)} \tag{4.35}$$

Again, both signs yield possible solutions. The negative sign is used for $\bar{T} < 1$. Once the flight velocity is found from Eq. 4.31, the rate of climb can then be obtained from Eq. 4.35. Of course, this process can also be reversed (i.e., the path angle can be determined) for a given $\bar{V}$ and T/W, from Eq. 4.33, whereupon the rate of climb is found from Eq. 4.35.

It should be emphasized again that Eqs. 4.33 and 4.35 represent general results for any value or expression of thrust. Eq. 4.31, however, is the result of using the simple thrust formula Eq. 3.33, which is independent of velocity.

EXAMPLE 4.4

The thrust of the aircraft in Example 4.3 is increased to 16,000 lb and it is climbing at a speed corresponding to E_m. Calculate the path angle and the rate of climb.

In this case

$$\bar{V} = 1$$

$$\bar{T} = \frac{T}{W}\frac{L}{D}\bigg|_{\max} = \frac{16{,}000}{36{,}000} \times 18.89 = 8.396$$

Eq. 4.33 is used:

$$\sin \gamma = \bar{V}^2 E_m - \sqrt{E_m^2 \bar{V}^4 - (2\bar{T}\bar{V}^2 - 1 - \bar{V}^4)}$$

$$= 18.89 - \sqrt{18.89^2 - (2 \times 8.396 - 3)} = 0.3956$$

Therefore

$$\gamma = 23.3°$$

The rate of climb is

$$v = V \sin\gamma = V_{D_{\min}} \sin\gamma = 356.6 \times 0.3956 \times 141 \frac{\text{ft}}{\text{sec}}$$

$$= 8466 \text{ fpm}$$

NOTE

The climb angle solution, and consequently the rate of climb, can be simplified and expressed in terms of level flight conditions for any angle and for unrestricted thrust expressions. To this end, the ratio of climb thrust to level flight thrust is formed by dividing Eq. 4.29 by Eq. 3.41, as follows:

$$\frac{T}{T_L} = \frac{\sin\gamma + \dfrac{1}{2(L/D)_{\max}}\left(\overline{V}^2 + \dfrac{n^2}{\overline{V}^2}\right)}{\dfrac{1}{2(L/D)_{\max}}\left(\overline{V}_L^2 + \dfrac{1}{\overline{V}_L^2}\right)} \tag{4.36}$$

where the subscript L refers to the level flight values at which the climb starts. It is also assumed that both climb and level flight occur at the same L/D. Recognizing that the denominator can also be expressed as $1/(L/D)$ (see Eq. 3.41), Eq. 4.36 becomes

$$\frac{T}{T_L} = \frac{L}{D}\sin\gamma + \frac{\overline{V}^2 + \dfrac{n^2}{\overline{V}^2}}{\overline{V}_L^2 + \dfrac{1}{\overline{V}_L^2}} \tag{4.37}$$

And since

$$n = \cos\gamma$$

$$\overline{V}^2 = \overline{V}_L^2 \cos\gamma$$

$$L = L_L \cos\gamma$$

Eq. 4.37 becomes

$$\frac{T}{T_L} = \frac{L}{D}\sin\gamma + \cos\gamma \tag{4.38}$$

Eq. 4.38 determines, for a given L/D and thrust, the climb angle γ. Conversely, for a given L/D and desired climb angle, the increase of thrust over the level flight value, T/T_L, can be established.

EXAMPLE 4.5

Solve Example 4.3 by using Eq. 4.38.

$$W = 36{,}000 \text{ lb}$$

$$T = 16{,}000 \text{ lb}$$

$$C_D = .014 + .05{C_L}^2$$

The aircraft in Example 4.3 is climbing at a speed corresponding to E_m or at $V_{D_{\min}}$. This is assumed to be also the original level flight speed.

At level flight T_L = D, and since at E_m

$$C_D = 2C_{D_0} = 2 \times 0.014 = 0.028$$

and

$$C_{D_0} = C_{D_i} = kC_L^2$$

then

$$C_L = \sqrt{\frac{0.014}{0.05}} = 0.529$$

The drag can be found from

$$D = W\frac{C_D}{C_L} = 36{,}000 \times \frac{0.028}{0.529} = 1905 \text{ lb} = T_L$$

Then

$$\frac{T}{T_L} = \frac{16{,}000}{1{,}905} = 8.4$$

and with

$$\frac{L}{D} = E_m = 18.89$$

one finds from Eq. 4.38 that

$$\gamma \approx 23^\circ$$

It is of interest to obtain two limiting cases from Eq. 4.10. For $\gamma = 0$ (level flight), Eq. 4.10 or 4.31 reduces to the basic situation of Eq. 3.48. For $\gamma = 90^\circ$ (vertical flight), one obtains from Eq. 4.31, with the negative sign, $\overline{V} = 0$. The positive sign gives

$$\overline{V}^2 = 2\left.\frac{L}{D}\right|_{\max}\left(\frac{T}{W} - 1\right) \tag{4.39}$$

Introducing the definition of $V_{D_{\min}}$ (see Eq. 4.10), and remembering that $E_m = 1/2\sqrt{kC_{D_0}}$, one finds after some rearrangement that

$$T - W = \frac{C_{D_0}\rho S V^2}{2} \tag{4.40}$$

or

$$T - W = D_{L=0} \tag{4.41}$$

which gives the condition for powered vertical climb if the thrust is independent of velocity. Eq. 4.41 states that, in vertical flight, the thrust must be at least equal to the aircraft weight, plus the zero lift drag. The implication here is that lift is zero for vertical flight and it is the thrust that contributes a balancing force to counter the weight W.

As a matter of completeness, it is possible to calculate from Eqs. 4.33 and 4.35 the maximum climb angle and the maximum rate of climb. However, considering the fact that the simplified thrust expression (Eq. 3.33) needs to be used for practical results and that the acceleration effect has been neglected ($\dot{V}$ term in Eq. 2.11), or the term in the denominator of Eq. 3.4, the slight improvement of results over those for the shallow flight path angle would hardly justify the extra labor involved.

4.4.3 Partial Power and Excess Power Considerations

In Section 4.1 it was shown that climb is the direct result of excess power available above and beyond the amount required to overcome aircraft drag. Both rate of climb and climb angle are determined by applied excess power. Since the climb starts usually from some steady-state level flight condition, it is of interest to reconsider what happens if steady-state power is increased, or how much power increase is required to achieve certain desired climb conditions. Moreover, what happens if the power is reduced?

In steady-state flight, each specific angle of attack gives a specific C_L with the corresponding EAS for a particular aircraft weight. An increase in angle of attack will produce a reduction in velocity, and a decrease in angle of attack will lead to an increase in velocity. During the change in angle of attack and while a new steady-state flight condition is established the aircraft may climb or descend if there is no change in power setting but the fundamental cause-effect relationship is between the angle of attack and the airspeed. Thus, one can conclude that the angle of attack is the primary control of the aircraft velocity in steady flight.

If the aircraft power is increased during steady-state flight conditions, the resulting excess power would cause a positive climb rate dh/dt, as already established in Section 4.2. An increase of power while maintaining the same angle of attack will result in aircraft climbing to a higher altitude. A decrease in power results in descent. Therefore, power setting is the primary control of altitude during steady-flight conditions.

The previous discussion has tacitly implied that typically "normal" relationships obtain between power applications and resulting flight speeds (i.e., a higher power setting gives a higher steady state flight speed, and vice versa). This is true if the flight speed is sufficiently high (higher than $V_{P_{\min}}$, or the speed for maximum endurance), but at speeds lower than $V_{P_{\min}}$ an increase in power is required as the flight speed decreases. This is seen in Figures 3.2 and 3.7. Since the increase in required power with a decrease in velocity is contrary to normal command of flight, the region between $V_{P_{\min}}$ and the stall speed V_s (minimum control speed) is called the region of reversed command, back side of power curve, or the slow flight region.

Consider an aircraft flying at point 1 (Figure 4.6) in the normal command (high speed) region. If the angle of attack is decreased without changing the power setting, a speed increased results with a deficiency in power and the aircraft will descend. If the angle of attack is increased, without change in power, the speed will decrease to point 2 and a climb will result (between points 1 and 2) due to an excess of power.

Flight in the reversed command means that a lower power setting is required at higher speeds, and vice versa, to maintain altitude. Thus, if aircraft angle of attack is increased at point 2 to produce an airspeed corresponding to point 3, power setting must be increased or descent would follow. Similarly, if the angle of attack is continuously decreased from point 3 to, say, $V_{P_{\min}}$, the increasing flight velocity permits a reduction in power if altitude is to be maintained. Otherwise, the aircraft would climb. Since the back side of the power curve is characterized by high lift coefficients, the reversed command operations occur during landing and take-off phases of flight. Most of the airplane flight takes place in the region of normal command.

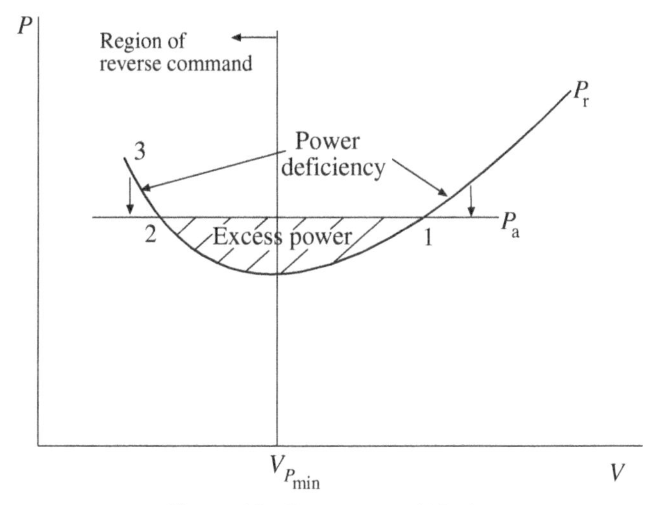

Figure 4.6 Power versus Velocity

PROBLEMS

4.1 Calculate the total power required (ft-lb)/sec for the rate of descent of 1,200 fpm and for the rate of climb of 1,200 fpm, for an aircraft in flight under the following conditions:

$$V_{\text{true}} = 300 \text{ ft/sec}$$
$$C_D = 0.05$$
$$S = 400 \text{ ft}^2$$
$$h = 10{,}000 \text{ ft}$$
$$\dot{W}_f = 5400 \text{ lb/hr}$$
$$W = 18{,}000 \text{ lb}$$

4.2 An aircraft weighing 20,000 lb begins a climb from sea level to 30,000 ft at an equivalent airspeed of 250 knots (423 ft/sec). This EAS will be maintained constant on the pilot's airspeed indicator throughout the climb. Assume a constant power available of 3,030 HP, and a constant average drag of 2,000 lb force. Calculate:

a. The TAS at the end of the climb (30,000 ft) and a mean TAS for the climb $(TAS_{30} + TAS_{\text{S.L.}})/2$

b. The power consumed in overcoming drag at the mean TAS, and the excess power available for the climb.

c. The resulting rate of climb (fpm), assuming constant TAS at the mean value (1,650 fpm).

d. The correction factor arising from acceleration during the climb, and the corrected mean rate of climb (fpm)

4.3 A P-3 has the following characteristics:

$$C_D = 0.0225 + 0.0448\ C_L^2 \text{ (clean)}$$
$$W = 110{,}000 \text{ lb}$$
$$S = 1300 \text{ ft}^2$$
$$h = 500 \text{ ft}$$
$$\eta_p = 0.82$$

It is cruising at maximum endurance speed (i.e., $V_{P_{\min}}$)
It is equipped with four Allison T56-A-14 turboprop engines.

What is its potential rate of climb if full power is applied?
Ans: 3910 fpm.

4.4 A jet aircraft is in steady, level flight at 20,000 ft altitude. It reduces thrust and descends to 15,000 ft altitude in 3.25 minutes. For the given aircraft data: $W = 24{,}000$ lb, $W/S = 60$, $V = 500$ mph (assume to be constant), $C_D = .025 + .06C_L^2$, determine:

a. The resulting path angle

b. The rate of sink

c. Horizontal distance covered

d. Maximum power-off glide range over the same altitude change

e. Flight speed corresponding to d.

4.5 For the turbojet aircraft with the characteristics given below, determine how much thrust must be developed at 300 knots to hold the rate of sink to 1,000 fpm at 10,000 ft. Consider thrust line inclination effects if applicable. The incidence angle of the wing is 2.5 with respect to the axis of the fuselage. The engine thrust line is along the axis. The $\alpha_{L=0}$ of the wing is along the axis of the fuselage.

$$C_D = 0.015 + 0.06\,C_L^2$$
$$W = 30{,}000 \text{ lb}$$
$$S = 750 \text{ ft}^2$$
$$a = 0.071 \text{ 1/deg}$$

4.6 A homemade glider has the following parameters: $W = 3{,}200$ lb, $C_D = .014 + .05C_L^2$. $S = 200\text{ft}^2$. It is easily established that the maximum glide range from 10,000 ft is 36 miles. How much thrust is needed from a retrofit jet engine to double this mileage? Ans: about 250 lb.

4.7 A high-performance aircraft cruises at 10,000 ft under minimum thrust conditions. The aircraft has the following characteristics:

$$W = 17{,}000 \text{ lb},\ S = 380 \text{ ft}^2$$
$$T = 14{,}000\,\sigma^{.8}$$
$$C_D = .022 + .0505C_L^2$$

Full available thrust is applied and the plane starts to climb at maximum climb angle. Determine:

a. The resulting climb angle at 10,000 ft altitude

b. The time to climb to 20,000 ft
Ans: 35.4°, about 66 sec.

4.8 Show that, for the same C_L, the power in climbing flight is given in terms of level flight power by

$$\frac{P}{P_L} = \cos^{3/2}\gamma\left(1 + \frac{L}{D}\tan\gamma\right)$$

Hint: First show that $V/V_L = \cos^{1/2}\gamma$.

4.9 Since maximum rate of climb is almost linear with altitude show that absolute ceiling h_c can be calculated from

$$h_c = \frac{h_1 v_o}{v_o - v_1}$$

where v_o and v_1 are maximum rates of climb at sea level and altitude h_1, respectively.

4.10 Using the results in Problem 4.9, calculate the absolute ceiling from the data tabulated in Example 4.5.
Ans: 19,020 ft.

4.11 If the maximum rate of climb varies linearly with the altitude show that the time to climb to altitude h is given by

$$t = \frac{h_c}{v_o}\ln\frac{h_c}{h_c - h}$$

4.12 Calculate the time to climb in Example 4.6 from the expression obtained in Problem 4.11.
Ans: 1,689 sec.

4.13 A propeller aircraft is climbing at the rate of 3,500 fpm with the following data:

$V = 190$ mph $\quad h =$ sea level
$W = 6{,}000$ lb $\quad b = 44$ ft
$S = 240\ \text{ft}^2$ $\quad f = 5.76$
$e = 0.91$ $\quad \eta_p = 0.85$

Determine the angle of climb and the P_r for the climb.

4.14 A test aircraft at 20,000 ft is flying straight and level at its minimum drag speed of 280 KTAS. Pilot then adds full power and commences a level acceleration run. Five seconds into the run, the airspeed has increased to 295 KTAS. Given the following aircraft parameters: $W = 40{,}000$ lb, $S = 400$ ft^2, $C_{D_0} = .025$, T_a is independent of V, determine:

a. Thrust available, T_a (*Hint:* For FPE first calculate average acceleration.)

b. V_{max}

c. Potential rate of climb at $V_{D_{min}}$

d. Maximum climb angle
Ans: $T_a = 9{,}046$ lb, $V_{max} = 1{,}195$ ft/sec, $R/C = 4{,}380$ fpm, 8.9°

4.15 A 24,000 lb jet aircraft in steady, level flight at a speed of 500 mph cuts its thrust back by 600 lb. $W/S = 60$, $C_D = .02 + .0535C_L^2$. Determine:

a. Resulting flight path angle

b. Its rate of sink

c. Horizontal distance covered when it reaches 15,000 ft
Ans: −1.43°, 18.33 ft/sec, 37.8 mi

4.16 A propeller aircraft establishes a steady-state climb at a constant equivalent airspeed of 140 knots. Passing through 5,000 ft the vertical speed indicator reads 600 fpm. Given: $SHP_{rated} = 300$ HP, $\eta_p = .85$, $W/S = 10$ lb/ft^2, $W = 3{,}000$ lb, $k = .0497$ $SHP_{avail} = (1.132\sigma - .132)$SHP.
Determine: P_{req}, f.

4.17 A jet with $S = 200$ft^2, $C_D = .014 + .05C_L^2$ is in a steady-state climb at 10,000 ft altitude. Determine:

a. The percentage increase in level flight thrust T_L required to achieve a climb angle of 30° if $V = V_{D_{min}}$.

b. If $T_{max} = 3T_{min}$, what is the climb angle, at the same airspeed?

c. If $V = 300$ ft/sec, $T/T_L = 7$, and the climb angle is 45°, how much does the aircraft weigh (this leads to a quadratic equation to solve)?
Ans: 930%, about 6°, 6,200 lb.

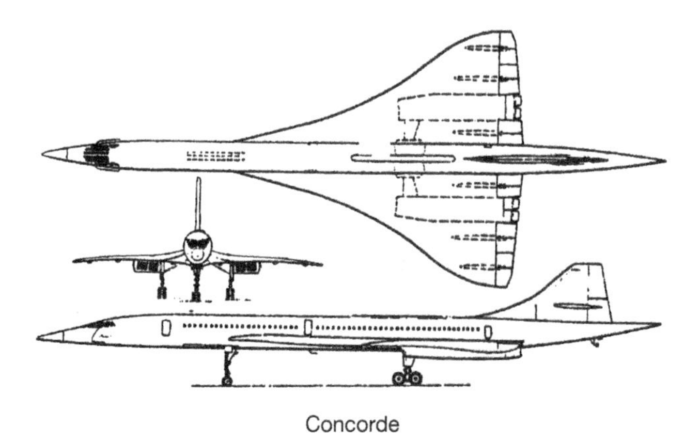

Concorde

5

Range and Endurance

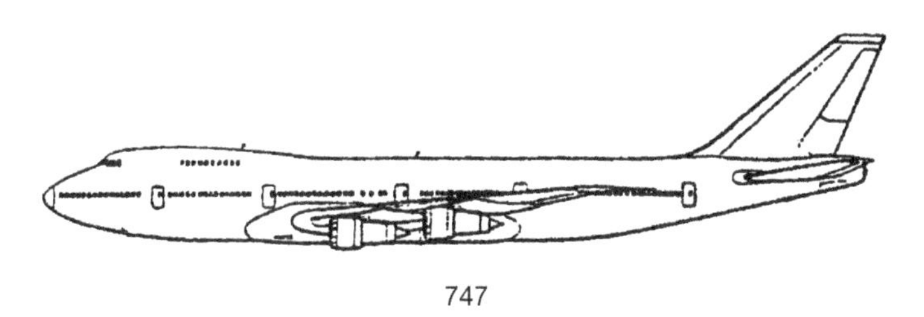

747

5.1 INTRODUCTION

Range is the distance covered on the ground while the aircraft is flying from one point in the space to another point. Endurance, at its basic level, may be defined as the time spent to cover that distance. However, an aircraft may spend four hours flying (theoretically) at 150 mph into a 150-mph headwind and not cover any distance at all. Thus, endurance needs to be defined in terms of flight time available, for a given amount of fuel, at a given flight condition.

Range calculation is subject to many conditions—in other words, it depends on the mission or flight profile. Different aircraft (or even the same aircraft) can cover the same distance in various ways resulting in different fuel consumption and endurance figures. The simplest mission would consist of take-off, cruise, and landing segments.

This chapter is concerned with the basic methods for calculating the aircraft range and endurance that would be applicable to any of the individual mission segments. The cruise range will be studied first in some detail, as the cruise conditions usually represent the aircraft design requirements and since most of the flying time is spent in the cruise configuration. The approach taken here is to use the same basic energy balance relationship, Eq. 2.20, as used for the majority of the problems treated in previous chapters. Thus, the change in energy level arising during the flight and its influence on the range can be correctly identified, and the influence of fuel and powerplant characteristics on the range can be studied separately. Eq. 2.20 can be rewritten by using Eq. 2.14:

$$W\,de = P_a dt - DV\,dt = \frac{\eta_o H_f}{g}\,dW_f - DV\,dt \tag{5.1}$$

Making use of the kinematic relationship Eq. 2.10, for level flight $dx \equiv dR = V\,dt$, one obtains from Eq. 5.1

$$dR = \frac{\eta_o H_f dW_f}{gD} - \frac{W\,de}{D} \tag{5.2}$$

Since also $dW_f = -dW$, $L = W$, and $D = W/(L/D)$ one gets

$$dR = \frac{-\eta_o H_f (L/D) dW}{gW} - \frac{L}{D}\,de \tag{5.3}$$

Eq. 5.3 is the general form for calculating the aircraft range. The basic requirements are that lift, drag, and weight variations be given as a function of two other flight parameters; velocity and altitude (see Section 2.1), or that some specific assumptions be introduced for approximate solutions. In particular, this leads to two different approaches: exact range integration—a numerical technique (Section 5.3), or a number of approximate methods (Section 5.2). The essential difference is that the range integration method uses a minimal number of assumptions but requires detailed thrust and drag data over applicable altitude and velocity ranges. This results in extensive numerical computations but gives highly accurate and reliable results. In practical applications, however, a number of simplifying assumptions are used to obtain usable but somewhat approximate equations and results. Several averaging or stepwise sequential calculations are then introduced to find adequate solutions.

Both approximate and integral methods are discussed in some detail. Initially, the term containing specific energy variation is neglected. Its effect on range (normally very small) is discussed in a later section. As already discussed in Chapter 3, separate solutions are obtained for propeller and jet aircraft.

5.2 APPROXIMATE, BUT MOST USED, METHODS

Before specializing Eq. 5.3 to particular flight paths or specific aircraft, it is possible to carry out a simple generalized integration if it is assumed that the lift-drag ratio and the overall power efficiency are constant. Then one obtains

$$R = \eta_o \left(\frac{H_f}{g}\right)\left(\frac{L}{D}\right) \ln \frac{W_1}{W_2} - \frac{L}{D}(e_1 - e_2) \tag{5.4}$$

which represents the aircraft range between points 1 and 2, provided that the energy assumptions discussed in Section 2.2 and $L = W$ are satisfied. It is easily recognized that the first term is due to change of weight (burning of the fuel) and the second term arises due to change in the specific energy level $(h + V^2/2g)$. For constant altitude and velocity cruise the second term vanishes. Its effect on range will be considered later in Section 5.4.2. Retaining, for the time being, only the first term in Eq. 5.4, one gets

$$R = \eta_o \left(\frac{H_f}{g}\right)\left(\frac{L}{D}\right) \ln \frac{W_1}{W_2} \tag{5.5}$$

which is a convenient form to observe the role of the fuel heat (energy) content in the range calculation process. The energy content H_f is a property of the fuel used, and η_o describes the ability of a particular engine to convert this energy into useful work. The lift/drag ratio L/D represents aircraft efficiency in conversion of the work into range. The role of the weight ratio W_1/W_2 is discussed further in Section 5.3. For hydrocarbons such as aviation gasoline a typical value for H_f is 18,800 BTU/lb (42,900 kJ/kg). If the heat capacity in BTU's is converted entirely by using the mechanical equivalent of heat, 778 ft-lb/

BTU (i.e., $\eta_o = 1$), it is seen that the remaining dimension is that of length

$$\eta_o \frac{H_f}{g} = \frac{g_c}{g} \frac{18{,}800 \left[\frac{\text{BTU}}{\text{lb}}\right] \times 778 \left[\frac{\text{ft-lb}}{\text{BTU}}\right]}{5{,}280 \left[\frac{\text{ft}}{\text{mi}}\right]} = 2770 \text{ miles} \quad (5.6)$$

H_f can be expressed in suitable units for calculation of the range. At first glance, the order of magnitude of the range can be estimated simply from the heat content of the fuel (implied here per lb of fuel). The influence of the other factors that determine the range is seen by assuming some typical values for a propeller driven aircraft as follows:

$$E_m = 12$$

$$\eta_o = 0.3$$

A 1 lb weight change in a 10,000 lb aircraft (i.e., $\ln(W_1/W_2) = 0.0001$) and assuming that fuel weighs 6 lb/gallon, then

$$\frac{R}{\text{gal}} = 0.3 \times 2{,}770 \times 6 \times 12 \times 0.0001 = 6 \frac{\text{mi}}{\text{gal}}$$

which is a typical value for reciprocating engines.

Although the last equation is not particularly useful for range calculation of specific aircraft it is suitable for comparative representation of aircraft performance. Individual aircraft range depends on particular power plant characteristics that are given in terms of specific fuel consumption. Thus, the previous equations must be further specialized for reciprocating and jet aircraft operations. In this chapter, following historical development, reciprocating engine will be treated first.

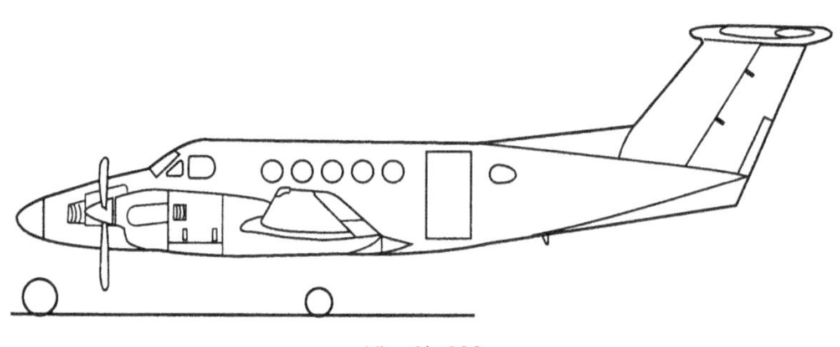

KingAir 200

5.2.1 Reciprocating Engine

In keeping with the energy approach used in Chapter 2, Eq. 2.14 can be cast in several different forms:

$$P_a = \frac{\eta_o H_f}{g}\frac{dW_f}{dt} = \frac{\eta_p \eta_t H_f}{g}\frac{dW_f}{dt} = \eta_p\, BHP \tag{5.7}$$

where

η_t = engine thermal efficiency
η_p = propulsive (propeller) efficiency
BHP = Brake horsepower

Since the specific fuel consumption can be expressed as

$$C = \frac{(dW_f/dt)}{BHP}\left[\frac{\text{lb}}{\text{HP} - \text{hr}}\right] \tag{5.8}$$

one obtains from Eqs. 5.7 and 5.8

$$\frac{\eta_o H_f}{g} = \frac{\eta_p\, BHP}{(dW_f/dt)} = \frac{\eta_p\, BHP}{C\, BHP} = \frac{\eta_p}{C} \tag{5.9}$$

Eq. 5.9 forms the basis for calculating the reciprocating aircraft range. Owing partly to historical developments and partly to calculational difficulties, two distinctive methods have emerged for calculating the reciprocating aircraft range. *Breguet method* was developed in the beginning of the twentieth century and is best known for its simplicity. The second one, having no specific label, could be called *constant velocity method.*

Breguet Equation The range, R, can be found directly upon substituting Eq. 5.9 and the conversion factor

$$\frac{550\left[\frac{\text{ft} - \text{lb}}{\text{HP} - \text{sec}}\right] \times 3{,}600\left[\frac{\text{sec}}{\text{hr}}\right]}{5{,}280\left[\frac{\text{ft}}{\text{mi}}\right]} = 375\left[\frac{\text{lb} - \text{mi}}{\text{HP} - \text{hr}}\right]$$

into Eq. 5.5 yielding the well known Breguet range equation for reciprocating engines. It should be remembered that the assumptions leading to Eq. 5.5 are η_p, C, and L/D are constant:

$$R = 375 \frac{\eta_p}{C} \frac{L}{D} \ln \frac{W_1}{W_2} \text{ [miles]} \tag{5.10}$$

For a typical heat content of $H_f = 18{,}500$ BTU/lb for aviation gasoline, the specific fuel consumption is

$$C \approx \frac{0.138}{\eta_t} \approx 0.45 - 0.55 \left[\frac{\text{lb}}{\text{HP} - \text{hr}}\right]$$

It is seen that, for constant η and C, the maximum range is obtained at E_m and at V_{E_m}. The velocity for maximum range can be obtained also from the power required velocity curve, Figures 3.4 or 5.1, by drawing a tangent to the curve from the origin. This is seen easily if one forms the tangent from P_r and V, whence

$$\tan\theta = \frac{P_r}{V} = \frac{\rho V^3 C_D S}{2V} \tag{5.11}$$

By eliminating $V^2 = 2W/\rho S C_L$ one gets

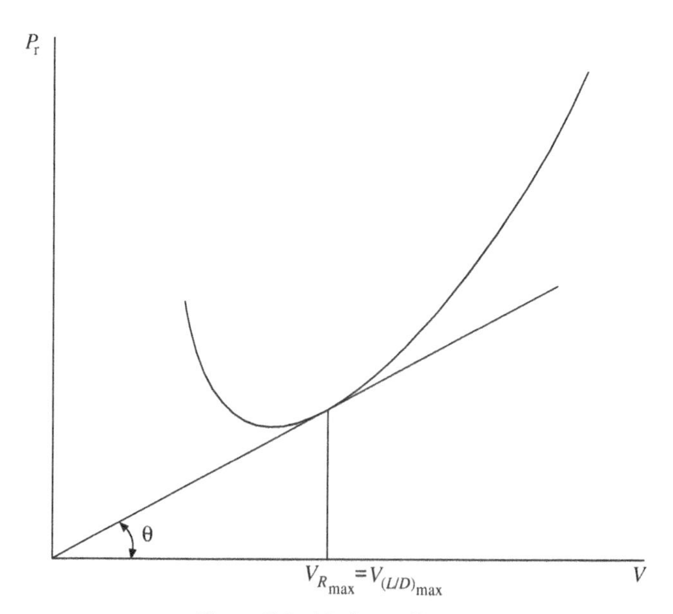

Figure 5.1 Maximum Range

$$\tan\,\theta = W\frac{C_D}{C_L} \approx \frac{1}{C_L/C_D} \tag{5.12}$$

Maximum value of C_L/C_D, or E_{m}, gives the minimum angle that is just tangent to the curve. With C_L = constant and with the flight velocity given by

$$V_{(L/D)\max} = \sqrt{\frac{2(W/S}{\rho}}\left(\frac{k}{D_{D_0}}\right)^{1/4} \tag{5.13}$$

the flight velocity needs to be continuously changed as the aircraft weight decreases. If the velocity is held constant, then W/ρ must also remain constant, which implies that as W decreases, the density must decrease. This implies that the aircraft will continuously drift up during the weight decrease. This is called cruise-climb schedule.

There are two problems with such a scheme. First, the air traffic control will have difficulties with the flights drifting all over the sky at varying altitudes and speeds. Thus, the Federal Aviation Administration (FAA), through the Federal Aviation Regulations (FAR) 21, 23, and 25, issues regulations for civil aviation concerning climb speeds, variation of cruise velocity, altitude variation, and so on. For example, even-numbered altitudes are assigned to traffic going in one direction and odd-numbered altitudes flying in opposite direction with altitude changes restricted to discrete steps. Military aviation is regulated by similar regulations through specific MILSPECS. Second, such a cruise-climb schedule may not be the best flight schedule for propeller aircraft, as the best propeller efficiency and the best fuel consumption cannot always be found at E_{m}.

Although the propeller efficiency and the specific fuel consumption C vary somewhat with power and rpm, they can be kept almost constant over the duration of flight. Since the Breguet equation tends to overestimate the range (the longer the range, the larger the discrepancy) there are two different approaches to Breguet equation for more accurate range prediction:

1. Mean, or average, quantities are used to establish both V and L/D.
2. Range is calculated from Eq. 5.10 in several short segments. This method is preferred in flight operations, as it yields a simultaneous fuel consumption check with the range schedule.

Constant Velocity Method Since the propeller aircraft tend to be flown at constant altitude and at constant throttle setting, a better range result can be found if one assumes a constant airspeed at constant altitude. From Eq. 5.3 and 5.9 one obtains

$$R = -375 \frac{\eta}{C} \int \frac{dW}{D} \tag{5.14}$$

Expressing the drag as

$$D = qSC_{D_0} + \frac{kW^2}{qS} \tag{5.15}$$

it follows that

$$\begin{aligned} R &= 375 \frac{\eta}{C} \left(\frac{1}{qSC_{D_0}} \int \frac{-dW}{1 + a^2W^2} \right) \quad \text{with} \quad a^2 = \frac{k}{q^2S^2C_{D_0}} \\ &= 375 \frac{\eta}{C} \left(\frac{1}{qSC_{D_0}a} [\arctan aW_1 - \arctan aW_2] \right) \end{aligned} \tag{5.16}$$

Eq. 5.16 can be simplified to read:

$$R = 750 \frac{\eta E_{\mathrm{m}}}{C} \left(\arctan 2C_{L_1} E_{\mathrm{m}} k - \arctan 2C_{L_1} E_{\mathrm{m}} k \frac{W_2}{W_1} \right) \tag{5.17}$$

Eq. 5.17 can also be expressed, as found in literature, as

$$R = 750 \frac{\eta E_{\mathrm{m}}}{C} \arctan \left[\frac{E_1(W_2/W_1)}{2E_{\mathrm{m}}(1 - kC_{L_1} E_1 W_2/W_1)} \right] \tag{5.18}$$

where

$$E_1 = \left.\frac{L}{D}\right|_1 \quad \text{and} \quad C_{L_1}$$

are determined at the beginning of the flight. Although this approach is better suited for general propeller aircraft analysis, the reason for the lack of its wide usage has been the difficulty of evaluation of the arctangent format with the tools prior to 1980s.

EXAMPLE 5.1

Determine the range of the following propeller-driven aircraft, with the following data, at a constant airspeed of 180 mph at 8,000 ft altitude:

$$W_1 = 18{,}500 \text{ lb}$$

$$W_{\text{fuel}} = 6{,}000 \text{ lb}$$

$$S = 939 \text{ ft}^2$$

$$\eta_p = 0.85$$

$$C = 0.45 \text{ lb/HP-hr}$$

$$C_D = 0.0192 + 0.047C_L^2$$

For use with Eq. 5.15 the following items need to be calculated

$$E_m = \frac{1}{2\sqrt{0.0192 \times 0.047}} = 16.64$$

$$C_{L_1} = \frac{2W}{\rho V^2 S}$$

$$= \frac{2 \times (18{,}500)}{(0.002377) \times (0.786) \times (264)^2 \times (939)} = 0.303$$

then

$$2 \times C_{L_1} \times E_m \times k = 2 \times .303 \times 16.64 \times .047 = .4739$$

and the range becomes

$$R = \frac{750 \times 16.64 \times .85}{.45} [\arctan .4739$$

$$- \arctan(.4739 \times (12{,}500/18.500)] = 3{,}123 \text{ miles}$$

For comparison, the range will also be evaluated by means of Breguet equation, Eq. 5.9. To this end, L/D will be evaluated at 180 mph:

$$C_L = \frac{2W}{\rho V^2 S} = \frac{2 \times 15{,}500}{(0.002377) \times (0.786) \times (264)^2 \times 939} = 0.254$$

where an average weight of 15,500 lb has been used. Then,

$$\frac{L}{D} = \frac{0.254}{0.0192 + 0.047 \times (0.254)^2} = 11.42$$

The range is obtained from Eq. 5.9 as

$$R = 375\frac{\eta}{C}\frac{L}{D}\ln\frac{W_1}{W_2} = 375 \times \frac{0.85}{0.45} \times 11.42 \times \ln\frac{18{,}500}{12{,}500}$$

$$= 3{,}171 \text{ miles}$$

Which shows unusually good agreement since the Breguet equation, in general, tends to overestimate the range by 10 to 15 percent. It should be noticed that an average weight and an $L/D|_{\text{mean}}$ were used in the last calculation. If, as usually assumed, one uses E_{m} in the Breguet equation, then the range becomes

$$R = 3{,}171\,\frac{16.64}{11.42} = 4{,}620 \text{ miles}$$

Thus, extreme care should be excercised in evaluating the common parameters in use of the Breguet equation.

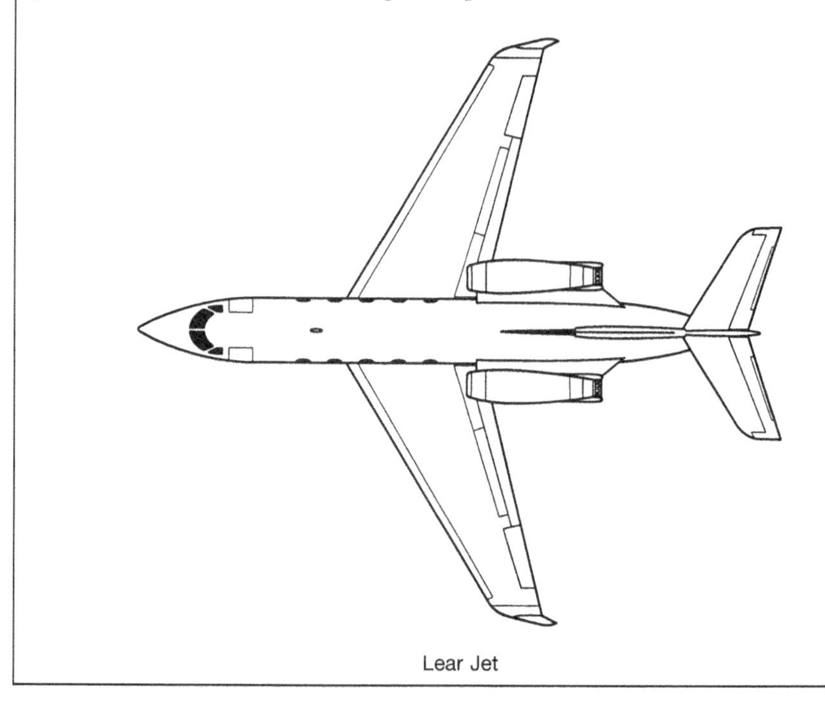

Lear Jet

5.2.2 Jet Aircraft

For a jet engine the available energy from the fuel can be expressed as

$$\eta_o H_f = \frac{P_a}{(dW_f/dt)} = \frac{TV}{TC} = \frac{V}{C} \tag{5.19}$$

with the specific fuel consumption C now defined as

$$C = \frac{\dot{W}_f}{T}\ [\mathrm{lb_m/hr/lb_f}] \tag{5.20}$$

Substituting now into Eq. 5.3, one obtains the following range integral:

$$R = -\int \frac{V}{C}\frac{dW}{D} - \int \frac{L}{D}\,de = -\int \frac{V}{C}\frac{L}{D}\frac{dW}{W} - \int \frac{L}{D}\,de \tag{5.21}$$

Similarly to propeller aircraft, simple analytical expressions can be found if a number of simplifying assumptions are made (i.e., aircraft cruise flight program will be prescribed). Three commonly used programs (sets of assumptions) will be discussed below.

Breguet Cruise—Climbing Flight The most commonly used expression for the jet engine range calculation is the Breguet analog for the jet aircraft. It is assumed that V = const, C_L = const, (L/D) = const (if the simple drag polar is used) and one obtains (neglecting again the second term in Eq. 5.21)

$$R = \frac{V}{C}\frac{L}{D}\ln\frac{W_1}{W_2}\ [\text{miles}] \tag{5.22}$$

where velocity is in mph. It is easily seen that, for maximum range with a constant C and a given weight ratio, $V(L/D)$ must be maximum. Thus, one finds that

$$V(L/D) = \sqrt{\frac{2W_1}{\rho S C_L}}\frac{C_L}{C_D} = \sqrt{\frac{2W_1}{\rho S}}\frac{\sqrt{C_L}}{C_D} \tag{5.23}$$

must be a maximum, or $C_D/\sqrt{C_L}$ must be a minimum, and one obtains with the parabolic drag polar the following minimum condition

$$\frac{d}{dC_L}\left(\frac{C_{D0} + kC_L^2}{\sqrt{C_L}}\right) = 0 \tag{5.24}$$

After carrying out the differentiation and simplifying, one finds

$$C_{D0} = 3kC_L^2 \tag{5.25}$$

and

$$C_{LR_{\max}} = \sqrt{\frac{C_{D0}}{3k}} = \frac{C_{LD_{\min}}}{\sqrt{3}}$$

It also follows that

$$V_{R_{\max}} = 3^{1/4}V_{D_{\min}} = 1.316V_{D_{\min}} = 1.316V_{E_{\mathrm{m}}} \tag{5.26}$$

One can also show that for maximum range

$$\frac{L}{D} = \frac{\sqrt{3}}{2}E_{\mathrm{m}}$$

which permits writing the maximum range expression from Eq. 5.22 as

$$R_{\max} = 1.14\frac{V_{D_{\min}}E_{\mathrm{m}}}{C}\ln\frac{W_1}{W_2} \tag{5.27}$$

Since Eq. 5.25 gives the lift coefficient for maximum range, the (equivalent) airspeed for the initial weight W_1 can be calculated. The appropriate cruise altitude can be obtained approximately from the thrust available data at a point where the thrust-altitude curve at cruising velocity $V_{R_{\max}}$ crosses the drag curve at EAS (see Figure 3.5). Similarly to the propeller-driven aircraft, the velocity for maximum range can be obtained from the thrust-velocity curve by drawing a tangent from the origin to the thrust-required curve.

The effect of weight change on the range can be assessed as follows. In the integration of Eq. 5.21 it was assumed that L/D = constant. If two weights are compared, at the same airspeed and constant C_L,

$$W_1 = \frac{1}{2}\rho_0\sigma_1 C_L V^2 S \tag{5.28}$$

$$W_2 = \frac{1}{2}\rho_0\sigma_2 C_L V^2 S \tag{5.29}$$

which gives

$$\frac{W_2}{W_1} = \frac{\sigma_2}{\sigma_1} \tag{5.30}$$

This shows that, as the aircraft weight decreases due to fuel being burned, the density ratio also must decrease. Thus, similar to the developments for propeller aircraft, as the fuel is expended the aircraft will gain altitude or will drift up and air traffic control issues are encountered again.

EXAMPLE 5.2

An F-86 aircraft has the following characteristics:

$C_D = 0.0159 + 0.075\ C_L^2$ (clean)

$W_1 = 14{,}200$ lb

$q = 192$ lb/ft^2

$S = 290$ ft^2

$W_f = 3{,}000$ lb

$\dot{W}_f = 1{,}000$ lb/hr

Calculate:

a. The range at constant velocity starting at 25,000 ft
b. The final altitude at the end of this range

a. For use with Eq. 5.22 C_L and C_D will be evaluated as follows:

$$C_L = \frac{W}{qS} = \frac{14{,}200}{192 \times 290} = 0.225$$

$$C_D = 0.0159 + 0.075 \times (0.255)^2 = 0.0208$$

which gives

$$\frac{L}{D} = \frac{0.255}{0.0208} = 12.26$$

TSFC is obtained by first determining drag

$$D = C_D qS = 0.0208 \times 192 \times 290 = 1{,}158 \text{ lb}$$

and then calculating

$$C = \frac{\dot{W}_\mathrm{f}}{T} = \frac{1{,}000}{1{,}158} = 0.863 \frac{\mathrm{lb_m}}{\mathrm{lb_f} - \mathrm{hr}}$$

The range is obtained as

$$R = \frac{V}{C}\frac{L}{D} \ln \frac{W_1}{W_2} = \frac{1}{C}\sqrt{\frac{2q}{\rho}}\frac{L}{D} \ln \frac{W_1}{W_2}$$

$$= \frac{0.6818}{0.863}\sqrt{\frac{2 \times 192}{0.002377 \times 0.448}}(12.26) \ln \frac{14{,}200}{11{,}200} = 1380 \text{ miles}$$

b. Final altitude is obtained by calculating the density ratio:

$$\sigma_2 = \sigma_1 \left(\frac{W_2}{W_1}\right) = 0.448 \left(\frac{11{,}200}{14{,}200}\right) = 0.353$$

which corresponds approximately to 32,000 ft.

Constant Altitude Cruise In addition to the constant altitude assumption, this program requires that either C_L = const and a parabolic drag polar exists or C_L = const and C_D = const. Eq. 5.21 can be written as

$$R = -\frac{(L/D)}{C}\int V\frac{dW}{W} = -\frac{(L/D)}{C}\sqrt{\frac{2}{\rho S C_L}}\int_1^c \frac{dW}{\sqrt{W}}$$

$$= 2\frac{(L/D)}{C}\sqrt{\frac{2}{\rho S C_L}}(\sqrt{W_1} - \sqrt{W_2}) = 2\frac{(L/D)}{C}V_1\left(1 - \sqrt{\frac{W_2}{W_1}}\right) \text{ miles} \tag{5.31}$$

where

$$V_1 = \sqrt{\frac{2W_1}{\rho S C_L}} \text{ mph}$$

is calculated at the beginning of the cruise. For calculation purposes, Eq. 5.31 is similar to Eq. 5.22, but in actual flight situations this flight program is not very practical since, as the weight changes, the flight velocity must continuously be reduced. As in the previous section, the maximum range is obtained at $(\sqrt{C_L}/C_D)_{max}$, which reduces to

$$C_{D_0} = 3kC_L^2 \tag{5.32}$$

as in Eq. 5.25.

EXAMPLE 5.3

Calculate the range for the aircraft in Example 5.2 for the constant altitude flight at 25,000 ft.

$$V_1 = \sqrt{\frac{2W_1}{\rho S C_L}} = \sqrt{\frac{2 \times 14{,}200}{0.002377 \times 0.498 \times 290 \times 0.255}}$$

$$= 570\ \frac{\text{ft}}{\text{sec}} = 388 \text{ mph}$$

The range is found from

$$R = \frac{2 \times 12.26}{0.863} \times 388 \times \left(1 - \sqrt{\frac{11{,}200}{14{,}200}}\right) = 1{,}234 \text{ miles}$$

Constant Altitude—Constant Velocity Cruise Since much of the cruise flight takes place at constant altitude and velocity, it is practical to consider flight with the assumptions that h and V are constant. In this case the range equation is written from Eq. 5.20:

$$R = -\frac{V}{C}\int \frac{dW}{D}$$

which is similar to the constant speed propeller range calculations, Eq. 5.18. Without going into the details, the range becomes

$$R = \frac{V}{C}\left(\frac{1}{qSC_{D0}}\int \frac{-dW}{1 + a^2W^2}\right) \tag{5.33}$$

with

$$a^2 = \frac{k}{q^2S^2C_{D0}}$$

Upon integration, the range can be expressed in two fully equivalent forms:

$$R = \frac{2V\,E_{\mathrm{m}}}{C}\left(\arctan 2kC_{L_1}\,E_{\mathrm{m}} - \arctan 2kC_{L_1}\,E_{\mathrm{m}}\,\frac{W_2}{W_1}\right) \tag{5.34}$$

or

$$R = \frac{2V\,E_{\mathrm{m}}}{C}\arctan\left[\frac{E_1(W_{\mathrm{f}}/W_1)}{2E_{\mathrm{m}}(1 - kC_{L_1}E_1W_{\mathrm{f}}/W_1)}\right] \tag{5.35}$$

where again

$$E_1 = \left.\frac{L}{D}\right|_1 \quad \text{and} \quad C_{L_1}$$

are evaluated at the beginning of flight.

This flight schedule is more practical, as it is able to conform with the general flight traffic control regulations. To maintain constant velocity, the thrust must be reduced as the aircraft loses weight due to the fuel usage. This can be carried out stepwise with the increase in the flight attitude.

5.3 RANGE INTEGRATION METHOD

A more accurate method for range calculation is a numerical-graphical integration technique. The advantage of this approach lies in the fact that it is independent of the type of power plant used. That is, it works equally well for reciprocating, turbojet, or turboprop engines. The only requirements are that the fuel flow rate, weight variation, and the true airspeed be given in a manner consistent with thrust or horsepower used. Thus, during the calculations airspeed and altitude may be varied continuously, which, in turn, determine the variations in thrust (or horsepower) required and the fuel used. This permits the calculation of the range under actual flight conditions, including the climb, cruise, and landing operations. Such calculations may be laborious but will provide highly reliable results, if so desired.

The starting point of the range calculation is the integral form of the range equation Eq. 5.21, which can be expressed (with the specific energy term neglected) as

$$R = \int_{W_1}^{W_2} \frac{V}{\dot{W}_\mathrm{f}}\, dW = \int_{W_1}^{W_2} \frac{VW}{\dot{W}_\mathrm{f}} \frac{dW}{W} \tag{5.36}$$

where

$$\dot{W}_\mathrm{f} \equiv \frac{dW_\mathrm{f}}{dt}$$

is the fuel flow rate. The quantity $V/\dot{W}_\mathrm{f}$ is called the specific range R_s and is a direct measure of the distance flown per unit fuel consumed. It is expressed as mi/lb, km/kg, or nautical air miles/lb (nam/lb). It is evident that the use of maximum values $V/\dot{W}_\mathrm{f}$ in Eq. 5.36 will also maximize the range. Since the fuel flow $\dot{W}_\mathrm{f}$, or the thrust specific fuel consumption (TSFC), is a function of altitude, velocity, and thrust, and since fuel flow rate is usually given in graphical form, determination of maximum values of specific range requires an optimization process that is best carried out by numerical-graphical means. Several variations of the specific range integration method can be found, depending on the user's purpose, approximations made concerning the altitude range, and the choice of dependent variable (h, V, M, or W). Three are discussed in the next section.

The first one contains the full method, as practiced in industry. The second version is contained in the example following the first one, which is a somewhat abbreviated version leading to practical results with a modicum of an effort. The basic idea here is to indicate how elementary path information can be extracted by user choice simplification. The third approach (an operational approach) simplifies the methodology as it would be applicable to mission analysis taking into account dropping the stores and refueling.

5.3.1 Basic Methodology

Version One The essential features, which are common to all variants, are given in the following step-by-step technique:

1. Select an altitude.
2. Select at least four flight gross weights to cover the expected range weight variations for each weight.
3. For each weight select at least four cruise velocities. At each cruise velocity determine the vehicle drag corresponding to selected weight, altitude, and speed conditions. Assuming that $T = D$.
4. Determine the net thrust per engine, T_n.
5. Determine for the required T_n, at selected velocity and altitude, the fuel flow rate $\dot{W}_f$ from the engine data. The total fuel flow rate is obtained from

$$\dot{W}_f = \dot{W}_{f_n} \times \text{\# of engines}$$

   From engine data find TSFC, then

$$\dot{W}_f = TSFC \times T$$

6. Calculate the specific range $R_s = V/\dot{W}_f$ for each of the cruise velocities selected in Step 3.
7. Repeat steps 2 through 6 and 1 through 6 for all desired weights and altitudes, respectively. The calculation process is facilitated by setting up Table 5.1.

The calculated results are plotted as shown in Figure 5.2.

The maximum range is obtained at the maximum values of R_s, for a chosen weight, that determine the velocity (or Mach number), which must be maintained to achieve the best range. The effect of cruise at constant velocity, or more important, at constant Mach number, is shown as a vertical line. The line identified as holding speed is com-

TABLE 5.1 Range Integration Method

h = constant						
Step	1	2	3	4	5	6
Variable	W	V or M	$T = D$	T_n	$\dot{W}_{fn}$	$R_s = \frac{V}{\dot{W}_f}$
Action	Select	Select	Calculate from drag polar	# of engines engine data	Determine from # of engines	Step $\frac{2}{5}$

monly established at a velocity corresponding to that at minimum drag (or slightly higher), and it occurs at a fuel flow slightly higher than the minimum flow.

The long-range cruise (LRC) line defines 99 percent of maximum range to account for about 1 percent reduction of R_s due to required altitude increase as the weight decreases during the flight. In practice, this altitude adjustment cannot be made continuously because it takes a finite time to burn up fuel and to achieve a new weight that requires that the thrust be increased to maintain flight at the proper altitude for a high R_s. Experience shows that the (stepwise) thrust increase leads to about 1 percent reduction in R_s.

The effect of weight change on altitude is shown in Figure 5.3, which represents the data calculated above and presented in Figure 5.2 for several altitudes when cross-plotted at constant W.

It is clearly seen that as W decreases, the altitude for maximum R_s increases with an increase in the optimum flight velocity. Although the effect of increasing altitude is to increase the specific range R_s, the

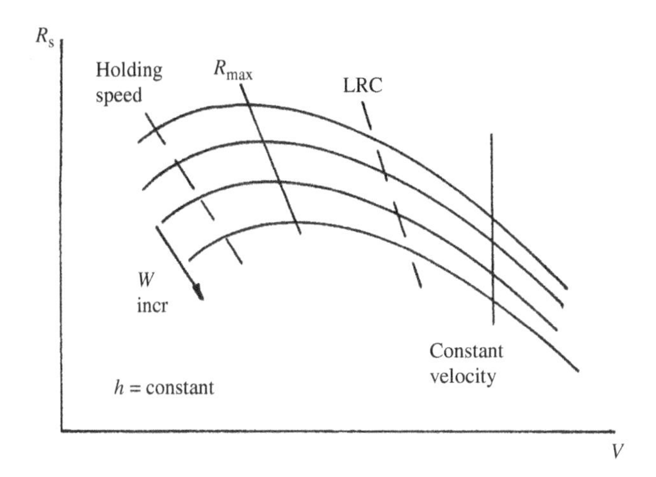

Figure 5.2 Specific Range at Constant Altitude

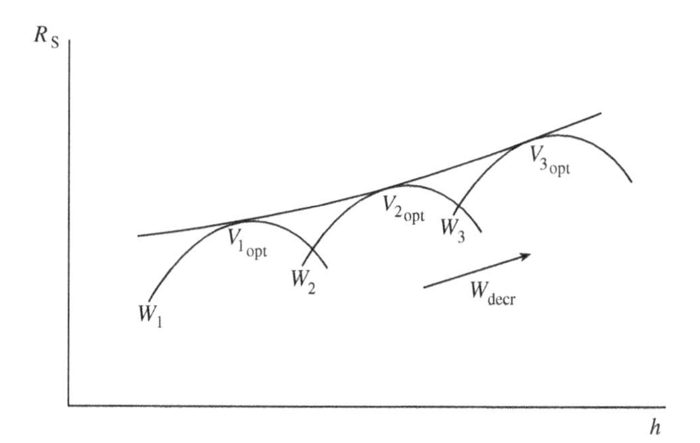

Figure 5.3 Effect of Weight and Altitude on Specific Range

thrust requirements (for increased altitude and velocity) lead to a decrease of R_s at higher altitudes. Thus, for each W there exists an optimum altitude, up to a point where engine thrust limitations or decrease in R_s restrict further altitude. The optimum altitude for each weight is determined if a quantity known as the range factor R_f, also called cruise constant, is plotted as a function of weight for a number of altitudes, as shown in Figure 5.4.

The range factor is defined as

$$R_f = \frac{VW}{\dot{W}_f} = W\,R_s = \frac{V}{C}\frac{L}{D} \tag{5.37}$$

which permits writing the range expression as

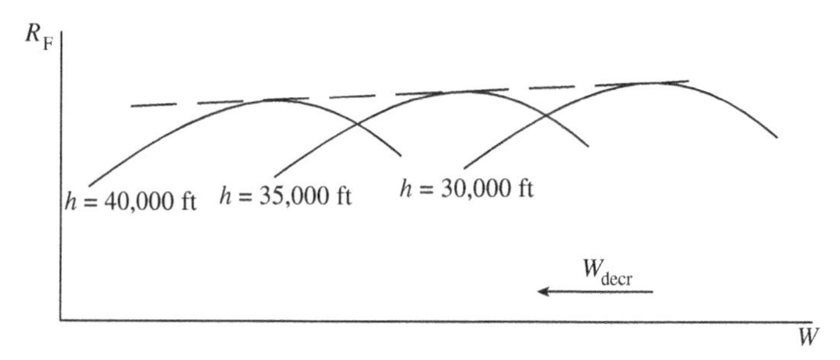

Figure 5.4 Optimum Altitude and Weight

$$R = R_f \ln\frac{W_1}{W_2}$$

(see Eqs. 5.36 and 5.22), and it is obtained from the data in Figure 5.2. It can be determined for maximum range, long-range cruise, or any other desired flight schedule. If plotted against weight W, the range factor often remains almost constant for maximum range (also LRC) for a climbing cruise schedule. Thus, the assumptions made for approximate integration of Eq. 5.21 are justified for long-range climbing cruise operation. Assuming the range factor to be a constant may not be valid for other range flight schedules.

Version Two This example illustrates a simplified version of the range integration method, which permits a quick evaluation of best cruise Mach number and cruise altitude. Obviously, these results can be sharpened by improved drag and thrust data.

EXAMPLE 5.3

For an eight-passenger corporate jet, determine the cruise Mach number, altitude, and fuel used for a range of 2,400 miles. The aircraft has the following characteristics:

$$W_{\text{cruise}} = 22{,}000 \text{ lb}, \; W_{\text{TO}} = 22{,}500 \text{ lb}, \; W_{\text{e}} = 16{,}000 \text{ lb (zero fuel)}$$

$$S = 300 \text{ ft}^2, \; AR = 7.7$$

$$T = 2 \times 2{,}100\, \sigma \cdot 55, \; h > 20{,}000 \text{ ft}$$

The drag data are given in tabular form for the simple drag polar:

M	.6	.7	.8	.9
C_{D_0}	.017	.017	.0175	.0195
k	.05	.05	.052	.055

Although the specific fuel consumption C varies with both altitude and Mach number, for simplicity only the variation with altitude is included here as C = .86, .88, .89 [$\text{lb}_\text{m}/\text{hr}/\text{lb}_\text{f}$], at 30,000, 40,000, and 50,000 ft altitude, respectively.

For calculations, it is assumed that the drag polar $C_D = C_{D_0} + kC_L^2$ holds, the weight is constant at 22,000 lb, and the range is calculated from Eq. 5.37. Thus, calculating only the range factor for selected Mach numbers (similar to Figure 5.2 will be

M	qS	C_L	C_D	D (lb)	T_a (lb)	$C\text{lb}_m/\text{hr}/\text{lb}_f$	R_f mi
.6	29700	.741	.0444	1320	1942	.88	7510
.7	45000	.543	.0318	1286	1942	.88	8965
.8	52800	.417	.0265	1401	1942	.88	9442
.9	66900	.329	.0255	1702	1942	.88	8708

sufficient for providing a selection process for the best Mach number and cruise altitude. The calculations are carried out in tabular format and are shown only for 40,000 ft altitude.

The range factor R_f is plotted in Figure 5.5 as a function of Mach number for 30,000, 40,000, 45,000, and 50,000 ft altitude. Best range is indicated by the highest value of the range factor. For simplicity, it will be assumed that the cruise Mach number corresponding to best range factor is the same as already given by the tabulated data. This permits a simple evaluation of the long-range cruise values at $.99R_f$. The following table, with liberal round-offs, summarizes the results.

h,kft	$M_{R_{max}}$	$R_f$mi	$.99R_f$mi	M_{LRC}
30	.7	8230	8150	.74
40	.8	9440	9350	.83
45	.82	9850	9750	.85
50	.89	10250	10000	.84

It may be concluded that the cruise Mach number should be in the .81 to .83 range. With the simplified drag polar and thrust values, ceiling is reached at about 50,000 ft and the realistic drag rise may be somewhat higher than used here. Thus, the long-range cruise Mach number is selected at a value less than .9. Also the specific fuel consumption C may be larger at Mach numbers higher than .8, thus further reducing the range factor. A compensating factor may be the use of constant weight W. In any case, this example shows the methodology, which may be sharpened by use of more detailed and complete thrust, drag, and weight data. The cruise fuel may be calculated from

$$\frac{W_1}{W_2} = \frac{22{,}000}{W_2} = \exp\left(\frac{2{,}400}{9{,}750}\right) = 1.279$$

and

$$W_f = 22{,}000\left(1 - \frac{1}{1.279}\right) = 4{,}800 \text{ lb}$$

5.3.2 An Operational Approach

From an operations point of view, range can also be evaluated if the specific range has been calculated for a particular cruise flight schedule by plotting R_s versus the weight, Figure 5.6. Since the data in the table above have been obtained for various weights, altitudes, and speeds, Figure 5.6 may represent any desired cruise schedule and is not limited to, say, constant altitude or velocity schedule.

The area under the curve, obtained by numerical integration between initial and final weights, is the desired range. W_1 is the initial weight ($W_2 + W_{fuel}$), and W_2 is the empty weight, (empty weight + fuel reserves).

The weight ratio W_1/W_2 has a strong influence on the range (see also Eqs. 5.9, or 5.21. On one hand, the lighter the aircraft (small W_2), the more range can be obtained for a given amount of fuel. On the other hand, increasing the dry weight W_2 requires a larger increase in

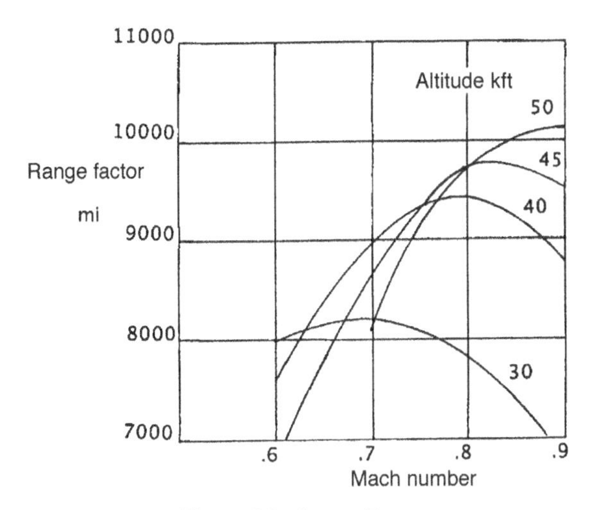

Figure 5.5 Range Factor

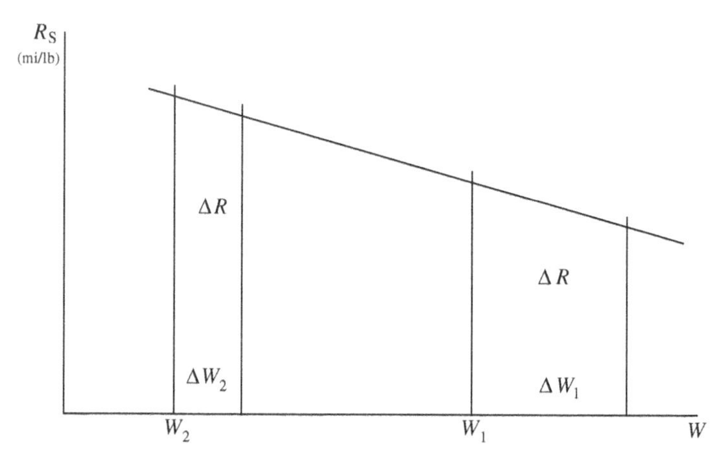

Figure 5.6 Range Solution

fuel weight if the range is to remain constant. In Figure 5.6 an empty weight increase ΔW_2 (due to equipment, stores, etc.) indicates that an area ΔR has been lost. For same desired range an equal ΔR must be made available by an increase of W_1 by ΔW_1. The primary reason for this disproportionate increase in fuel weight is the much higher specific range at the end of the flight (near W_2) where lower aircraft weight requires less thrust power. For convenience, the results of integration of the Figure 5.6 can be displayed as a plot of range against the instantaneous weight, Figure 5.7, which permits a quick evaluation of the available range at any given weight and is a valuable tool in mission analysis.

For practical calculations and mission analysis, the range integration process just described can be simplified to a few essential steps:

1. Fix initial aircraft weight W and altitude (ρ).
2. Choose V, or calculate $V_{R_{\max}}$.

   Jet Aircraft

$$V_{R_{\max}} = 1.316 V_{D_{\min}} = 1.316 V_{(L/D)_{\max}} = 1.316 \sqrt{\frac{2W}{\rho S}} \left(\frac{k}{C_{D_0}}\right)^{1/4} \tag{5.38}$$

   Propeller Aircraft

$$V_{R_{\max}} = V_{(L/D)_{\max}} = \sqrt{\frac{2W}{\rho S}} \left(\frac{k}{C_{D_0}}\right)^{1/4} \tag{5.39}$$

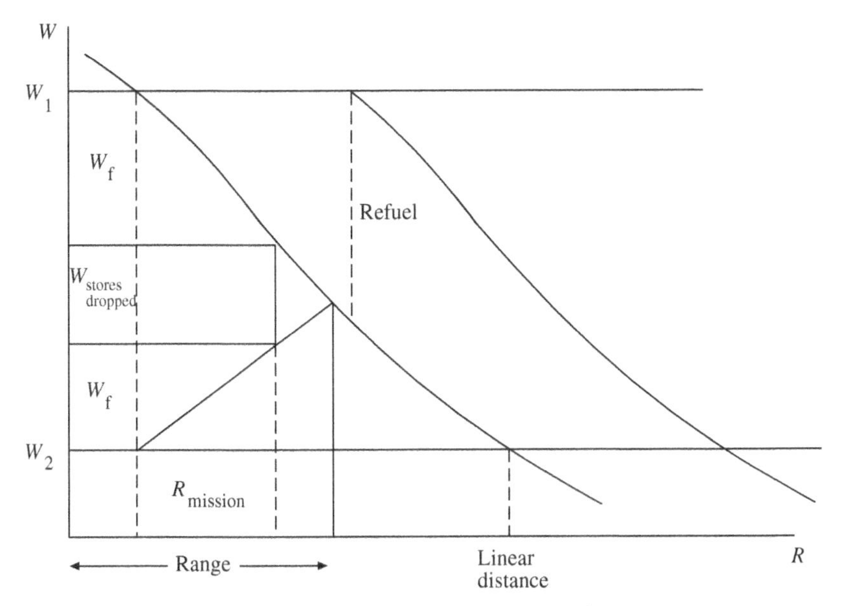

Figure 5.7 Mission Range (Radius)

or, the maximum range velocity can be obtained by drawing a tangent to the P_r or T_r curves (at a suitable W) as discussed in Section 5.1.

3. Calculate T_r or P_r from Eq. 5.15 or 3.55, or determine from T_r or P_r curves.
4. From manufacturer's data, determine the fuel flow rate $\dot{W}_f$ (lb/hr).
5. Calculate the specific range R_s:

$$R_s = \frac{V}{\dot{W}_f}\left[\frac{\text{mi/hr}}{\#\text{ of engines} \times \text{lb/hr}}\right] = \left[\frac{\text{mi}}{\text{lb}}\right]$$

6. Plot R_s vs. W.
7. Repeat steps 1 through 6 by varying W and σ as desired.
8. Integrate the $R_s - W$ curve to get instantaneous distance as a function of aircraft instantaneous weight.

Figure 5.7 shows the instantaneous weight as a function of downstream distance. At W_1 the distance covered is zero; full range is found at W_2. Such a graph may be prepared for any flight schedule (varying altitude, velocity, etc.). For most of the staightforward flight schedules, the curve is almost a linear one, and the radius (flight out and return) can be established simply by reversing the curve at the same slope to end at the empty weight W_2, as shown in Figure 5.7. Also shown are the effects of dropping the stores (bombs, fire retardant fluid, agricultural sprays, etc.) and refueling. The latter case is shown with the assumption that the aircraft is refueled back up to its original full weight, W_1. This may or may not be possible, depending on the configuration of volume available for fuel in case of dropped stores.

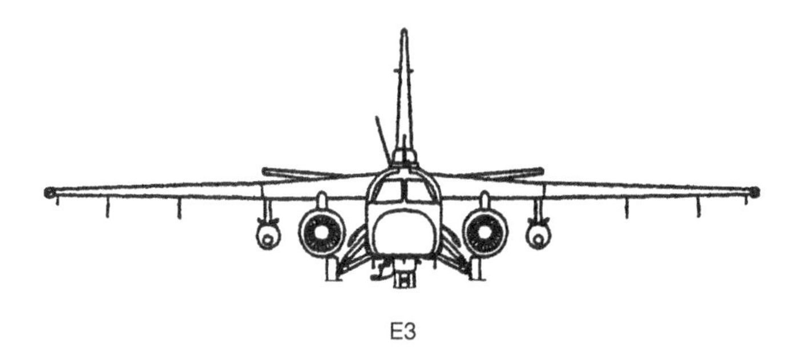

E3

5.4 OTHER CONSIDERATIONS

5.4.1 Flight Speeds

The range equations developed in Sections 5.1 and 5.2 are, at best, good approximations of what can be expected under actual flight conditions from take-off to landing. In addition to assumptions necessary for reaching useful solutions, air traffic control, weather, and variations in flight speed contribute to deviations from expected results. Similarly, the corresponding flight velocities (e.g., Eqs. 5.38 and 5.39), deduced while establishing the range expressions, also represent good guidelines and have long been subject to discussions as to what is the right or best speed to achieve a desired/maximum range.

For jet aircraft, it is intuitive to assume that a high flight speed is called for a long range. Indeed, Eq. 5.38 shows that high flight velocities are to be expected. For high wing loading (W/S) and high altitude (low ρ), the calculated maximum range velocity $V_{R_{max}}$ can approach and even exceed the sonic speed. This invalidates the assumptions of constant C_{D_0} and k, which can double and be reduced by a factor of two to three, respectively, beyond the critical Mach number range (see also Appendix B). Introduction of the local compressible flight regime values for C_{D_0} and k will help solve this problem and yield Eq. 5.38 as a reasonable estimate of the flight velocity. It should be noted that Eq. 5.38 also implies that the flight take place at $\sqrt{C_L}/C_D|_{\max}$.

For propeller aircraft, where at lower wing loadings and at lower altitude C_{D0} and k remain constant, Eq. 5.39 tends to predict flight velocities that are too low for practical flight purpose. The same conclusions can be drawn from Eq. 5.16, which predicts optimum flight speeds in the region of reversed command ($V < V_{E_m}$). Obviously, this contradicts the reasons for flight (high speeds at reasonable economy), as now the flight is supposed to take place at low speed and higher power (higher fuel consumption).

To reconcile the desire for higher speed and better economy, Carson (see bibliography) proposed, by reviewing the small aircraft performance, that minimizing the weight of fuel used per unit of velocity leads to a higher and more economical flight speed. Consequently, the equation for preferred flight speed turns out to be that given by Eq. 5.38. Thus, the expression for speed for desired economy for small aircraft and the best range speed for jet aircraft coincide. This does not mean that a small propeller-driven aircraft be flown at the jet speed. It means that Eq. 5.38 be used, with the values appropriate for a small aircraft, for determining the flight speed. As a result, the flight velocity turns out to be higher than given by Eq. 5.39 but appreciably lower than the speed obtained for a jet aircraft from Eq. 5.38.

This somewhat surprising conclusion is also available from the energy-based range equation, Eq. 5.2. For jet aircraft, Eq. 5.21 provides

$$dR = -\frac{V}{\dot{W}}\,dW = -\frac{1}{C}\frac{V}{T}\,dW$$

and maximizing

$$\frac{V}{T} = \frac{\sqrt{C_L}}{C_D}\sqrt{\frac{2}{\rho WS}}$$

maximizes the range with the best range speed given by Eq. 5.38 (see also Problem 5.7).

For propeller aircraft (see Eq. 5.9)

$$dR = -\frac{\eta_o H_f}{g}\frac{dW}{D} = -\frac{\eta_p BHP}{D\dot{W}}\,dW = -\frac{\eta_p}{CT}\,dW = -\frac{\eta_p}{C}\frac{L}{D}\frac{dW}{W}$$

which clearly implies that the best range is obtained at V_{E_m} (Eq. 5.39). Rewriting this now as

$$\frac{dW}{V} = -\frac{C}{\eta_p}\frac{T}{V}\,dR$$

it is easily seen that the Carson economy is obtained at maximum V/T, implying that the best speed then is at $V = 1.316V_{E_m}$.

5.4.2 Effect of Energy Change on Range

In Section 5.2 it was assumed that the specific energy $(h + V^2/2g)$ remains constant during the cruise portion of the flight. Consequently, the second term in the range equation vanishes. However, in deriving the range equation (e.g., Eq. 5.4), it was assumed that L/D is constant, which, in turn, implies that aircraft attitude is constant. Subsequent sections have shown that either the velocity or altitude must then change, with a resulting change in energy level. Usually, velocity is kept constant, and altitude variation brings about the resulting changes.

In order to estimate the magnitude of the omitted term, Rutowski (see bibliography) replaced specific energy variation de with altitude dh and developed from the hydrostatic equation

$$dh = -\frac{dW}{g\rho} = -\frac{RT}{g}\frac{dp}{p} = -\frac{RT}{g}\frac{dW}{W}$$

where

$$\frac{dW}{W} = \frac{dp}{p}$$

is obtained from logarithmic differentiation of the lift equation

$$L = W = \frac{\gamma}{2}\,pM^2SC_L$$

Thus, the energy equation may be written for a jet aircraft as

$$R = \left(\frac{V}{C} - \frac{RT}{g}\right)\frac{L}{D}\ln\frac{W_1}{W_2}$$

For a typical jet aircraft in stratosphere (T = 390R) at 800 ft/sec and with a specific fuel consumption of C = 1.0 $\mathrm{lb_m/lb_f/hr}$, the second term amounts to about .8 percent correction.

Since weight change is the basic cause of change(s) in energy, an approximate assessment of this effect can be made by rewriting Eq. 5.4 and Eq. 5.21 as

$$R = \frac{L}{D}\left[\frac{V}{C}\ln\frac{W_2}{W_1} - \left(\frac{E}{W}\bigg|_2 - \frac{E}{W}\bigg|_1\right)\right]$$

Assuming now that E = constant, which implies a cruise-climb operation at constant Mach number in the troposphere, then the energy term can be expressed as

$$E\left(\frac{1}{W_2} - \frac{1}{W_1}\right) = \frac{E}{W_1}\left[\frac{f}{1 - f}\right]$$

where f is the fuel fraction $f = W_f/W_1$. Then the range equation becomes

$$R = \frac{L}{D}\left[\frac{V}{C}\ln\frac{W_2}{W_1} - e_1\frac{f}{1 - f}\right]$$

For an aircraft traveling at a speed of 800 ft/sec at 40,000 ft, with a weight of 400,000 lb and a specific fuel consumption of $C = .6\ \text{lb}_m/\text{lb}_f/\text{hr}$, the relative magnitude of the second term is about 1.2 percent of the Breguet range—the first term.

5.5 ENDURANCE

Endurance is the time an airplane is able to remain in flight without landing. In the simplified analysis to follow this means also without refueling. Endurance with refueling can be included along the ideas outlined in the previous sections. Endurance can be obtained from the range equation by replacing $dR = Vdt$. Thus, Eq. 5.3 becomes

$$\begin{aligned} dt &= \eta_o\left(\frac{H_f}{g}\right)\left(\frac{L}{D}\right)\frac{dW}{VW} - \frac{(L/D)}{V}de \\ &= \eta_o\left(\frac{H_f}{g}\right)\left(\frac{L}{D}\right)\frac{dW}{VW} - \frac{(L/D}{V}dh - \frac{(L/D)}{g}dV \end{aligned} \tag{5.40}$$

where the specific energy has been used to expand the last term. For most ordinary applications, the last two terms can be neglected. Again, the nature of useful work done $\eta_o H_f$ leads to different expressions for the reciprocating and jet engine endurance calculations.

5.5.1 Reciprocating Engines

Using Eq. 5.8 one obtains

$$dt = -\frac{\eta}{C}\frac{L}{D}\frac{dW}{VW} \tag{5.41}$$

Since the weight changes, with L/D = constant, also the aircraft speed changes. Substituting the expression for the velocity

$$V = \sqrt{\frac{2W}{\rho S C_L}}$$

Eq. 5.41 becomes

$$dt = -\frac{\eta}{C}\sqrt{\frac{\rho S}{2}}\frac{C_L^{3/2}}{C_D}\frac{dW}{W^{3/2}} \tag{5.42}$$

Assuming now that the flight occurs at constant altitude, and that η, C, $C_L^{3/2}/C_D$, and L/D are constant, then the integration can be carried out to give

$$\begin{aligned} E = t_2 - t_1 &= \frac{2\eta}{C}\sqrt{\frac{\rho S}{2}}\frac{C_L^{3/2}}{C_D}\left(\frac{1}{\sqrt{W_2}} - \frac{1}{\sqrt{W_1}}\right) \\ &= \frac{2\eta}{C}\sqrt{\frac{\rho S C_L}{2W_1}}\frac{C_L}{C_D}\left(\sqrt{\frac{W_1}{W_2}} - 1\right) \\ &= \frac{2\eta}{C}\frac{C_L/C_D}{V_1}\left(\sqrt{\frac{W_1}{W_2}} - 1\right) \end{aligned} \tag{5.43}$$

Maximum endurance is obtained if the multiplier

$$\frac{\eta}{C}\sqrt{\frac{2\rho S}{W}}\frac{C_L^{3/2}}{C_D}$$

is maximum. For constant-altitude flight, the maximum occurs when $C_L^{3/2}/C_D$ is maximum. This condition can easily be found for the parabolic drag polar if the following differentiation is carried out:

$$\frac{d}{dC_L}\left(\frac{C_L^{3/2}}{C_{D_0} + kC_L^2}\right) = 0$$

Upon differentiation and simplification, one obtains

$$3C_{D_0} = kC_L^2 \quad \text{or} \quad C_D = 4C_{D_0}$$

But this is the condition for minimum power, as already established in Section 3.3.3. It occurs at

$$V_{P_{\min}} = \sqrt{\frac{2W}{\rho S}}\left(\frac{k}{3C_{D_0}}\right)^{1/4}$$

Then also

$$\frac{C_L}{C_D} = \frac{1}{4C_{D_0}}\sqrt{\frac{3C_{D_0}}{k}} = \frac{\sqrt{3}}{4\sqrt{kC_{D_0}}}$$

is constant, and $C_L^{3/2}/C_D$ is given by

$$\frac{C_L^{3/2}}{C_D} = \frac{3^{3/4}}{4k^{3/4}C_{D_0}^{1/4}}$$

As the aircraft weight W changes during the flight, the best speed for maximum endurance $V_{P_{\min}}$ decreases (see Figure 3.7 with Example 3.4), thus requiring that the aircraft speed be continuously changed during the flight. Since V_2 (speed at the end of flight) will be less than V_1, the specific energy term indicates a slight improvement in the endurance. In practice, however, such continuous speed change is impractical to maintain. Also, the best propeller efficiency η and the best fuel consumption C cannot always be maintained. Eq. 5.43 shows that the propeller airplane endurance is maximum at sea level. It can be written in a consistent set of units as follows:

$$E = 1{,}100\,\frac{\eta}{C}\sqrt{\frac{\rho S}{2W_1}}\,\frac{C_L^{3/2}}{C_D}\left(\sqrt{\frac{W_1}{W_2}} - 1\right) = 1{,}100\,\frac{\eta}{C}\,\frac{C_L}{C_D}\,\frac{1}{V_1}\left(\sqrt{\frac{W_1}{W_2}} - 1\right) \tag{5.44}$$

For the case of a parabolic drag polar the maximum propeller aircraft endurance can be expressed as

$$E = \frac{628\eta\rho}{Ck^{3/4}C_{D0}^{1/4}}\sqrt{\frac{\rho S}{2W_1}}\left(\sqrt{\frac{W_1}{W_2}} - 1\right) \tag{5.45}$$

In Eqs. 5.44 and 5.45

E—Hours $\quad$ η—Propeller efficiency
W—lb $\quad$ V—mph
ρ—slugs/ft^3 $\quad$ C—$\dfrac{\text{lb}}{\text{BHP} - \text{hr}}$

Subscripts 1 and 2 indicate quantities at the beginning and end of flight, respectively.

5.5.2 Turbojets

The endurance of a turbojet aircraft can be obtained also from Eq. 5.40 with the fuel energy expression used in Section 5.2.2.

$$\eta_o\left(\frac{H_f}{g}\right) = \frac{V}{C}$$

Thus, one obtains, with the energy term omitted, and assuming that L/D is approximately constant

$$dt = -\frac{L/D}{C}\,\frac{dW}{W} \tag{5.46}$$

and upon integrating

$$E = \frac{L/D}{C} \ln \frac{W_1}{W_2} \tag{5.47}$$

It should be noticed that, with the assumptions used, jet aircraft endurance could be obtained directly from the range expression, Eq. 5.22, by dividing the range by V. Eq. 5.47 shows that the jet aircraft maximum endurance occurs at E_m (with the fuel consumption C being constant). For a parabolic polar this condition was shown in Section 3.3.2 to give $E_m = 1/(2\sqrt{kC_{D0}})$ and it occurs at the velocity for minimum drag

$$V_{D_{min}} = \sqrt{\frac{2W}{\rho S}} \left(\frac{k}{C_{D0}}\right)^{1/4} \tag{5.48}$$

Thus, the endurance can be rewritten for parabolic polar:

$$E = \frac{1}{2C\sqrt{kC_{D0}}} \ln \frac{W_1}{W_2} = \frac{E_m}{C} \ln \frac{W_1}{W_2} \tag{5.49}$$

$V_{D_{min}}$ shows that the best endurance speed is dependent on the aircraft weight and decreases as the flight progresses (see Figure 3.4 with Example 3.3). According to Eq. 5.49, the endurance does not seem to be dependent on the altitude. However, the specific fuel consumption C tends to increase with increasing altitude and increasing thrust. Thus, it is expected that improved endurance can be expected at lower altitudes at modest thrust levels.

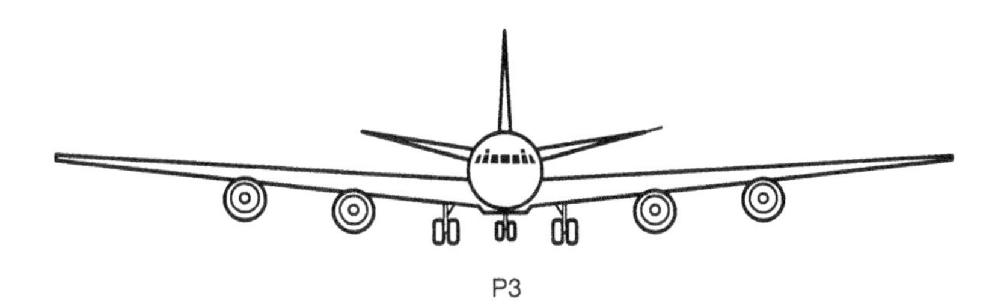

P3

5.5.3 Endurance Integration Method*

Similarly to the range integration method discussed in Section 5.2, aircraft endurance can also be evaluated for any flight path with good accuracy from

$$E = \int_{W_2}^{W_1} \frac{dW}{\dot{W}_f}$$

For jet aircraft

$$\frac{1}{\dot{W}_f} = \frac{R_s}{V}$$

can be obtained directly from tabulated range calculations in Section 5.5. The propeller aircraft fuel flow rate is evaluated as

$$\dot{W}_f = \frac{CDV}{\eta \rho}$$

as seen from Eq. 5.41. The calculation and graphing procedure for $1/\dot{W}_f$ as a function of altitude and weight is the same as for the range, Figure 5.8, and needs not be repeated here again.

5.6 ADDITIONAL RANGE AND ENDURANCE TOPICS

The previous sections of this chapter have clearly shown that cruise range and endurance evaluation is simple and straightforward if commonly accepted special conditions can be shown to be valid. The numerical (range) integration scheme hinted of the complexities involved if more exact results or noncruise flight schedule results are desired. Indeed, accurate range calculations under a variety of conditions and

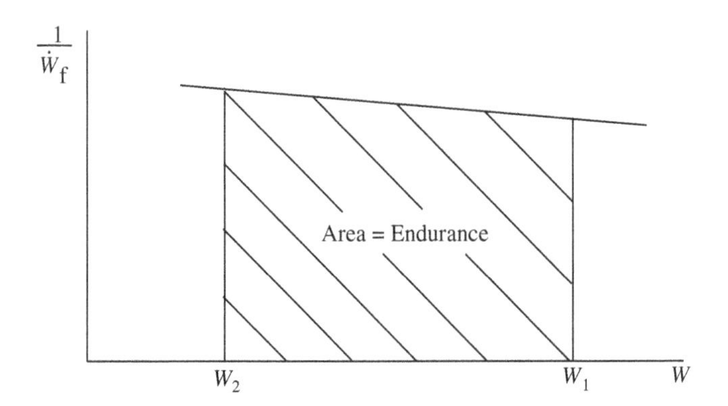

Figure 5.8 Endurance Calculation

factors involved are important for any successful commercial operation. For military and special-purpose airplanes, range or endurance calculations depend on type of mission flown, refueling, stores dropped during flight, and so on. In the following sections the effect of wind on range and endurance will be considered as it influences all other results, regardless of the flight schedules. Also, some general comparisons of range and endurance results will be offered.

5.6.1 The Effect of Wind

Up to this point, the basic performance analyses were based on the assumption that the flight path was determined by lift and thrust and not subject to wind disturbances. In practical operations, such a condition would be the exception rather than the rule. The wind can have a strong influence, both adverse and beneficial, on aircraft performance: range, take-off, landing, and so on. However, wind has no practical influence on endurance. To illustrate this, consider the T_r versus V curve in Figure 5.9.

The same argument holds for the power-required curve of propeller-driven airplanes. The presence of wind does not change the shape of the curve. Only the locus of the range velocity on the curve (the amount of drag) is modified by wind. If the airplane is flying, say, into the

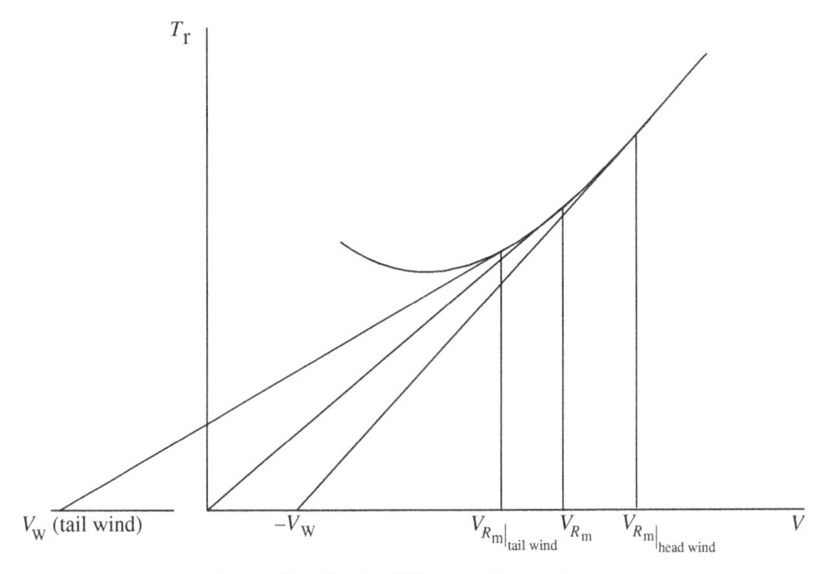

Figure 5.9 Effect of Wind on Range Speeds

wind, its relative airspeed, V_w, is increased but its ground speed, V_g, is decreased:

$$V_g = V - V_w \text{ (head wind)} \tag{5.50}$$

$$V_g = V + V_w \text{ (tail wind)} \tag{5.51}$$

As the range is defined by the ground speed, it is decreased by the head wind. In Figure 5.9 this amounts to moving the origin to the right. Drawing a tangent to the thrust-required curve locates the maximum range speed for the flight with the head wind. The tail wind shifts the origin to the left. If the plane flies into the head wind with the airspeed equal to the wind speed V_w, the ground speed and the range become zero. This flight, at zero range, can be continued (theoretically) until the aircraft runs out of fuel.

The zero range finite endurance flight is consistent with the definitions of endurance and range. Range is dependent on ground speed while endurance is determined by the fuel available for propulsion. It should be pointed out that the specific fuel consumption C is affected by flight speed and the thrust level (drag). However, once the lowest drag is established on the thrust curve, then also the best endurance relative air speed is determined and the endurance depends only on the amount of fuel available. Rather than work with the elusive tangents in Figure 5.9 it is possible to formulate the previous concepts in analytical form. Propeller and jet powered aircraft will be treated separately.

Propeller Aircraft To properly formulate the range expression with the wind effect, included it is best to return to the basic range energy equation, Eq. 5.1, and to identify the velocities as follows:

$$W\,de = \eta_o H_f dW_f - DV_a dt \tag{5.52}$$

and

$$dR = V_g dt \tag{5.53}$$

where

V_a = airspeed
V_g = ground speed

Combining Eq. 5.52 and Eq. 5.53 gives

$$dR = -\frac{\eta}{C}\frac{V_{g}}{V_{a}}\frac{dW}{D} - \frac{V_{g}}{V_{a}}\frac{W}{D}\,de \tag{5.54}$$

With the assumptions of $de \approx 0$, C_L = const and V = const, the range equation becomes

$$R = 375\,\frac{\eta}{C}\frac{L}{D}\left(1 \pm \frac{V_{w}}{V_{a}}\right)\ln\frac{W_1}{W_2} \tag{5.55}$$

If the velocity is nondimensionalized by use of $V_{D_{\min}}$ (Eq. 3.26), and Eq. 3.36 is introduced in the following form:

$$\frac{L}{D} = \left(\frac{2\overline{V}^2}{\overline{V}^4 + 1}\right)E_{m}$$

then the range equation with the wind can be written as

$$R = 375\,\frac{\eta}{C}\,E_{m}\,(\overline{V} \pm \overline{V}_{w})\left(\frac{2\overline{V}}{\overline{V}^4 + 1}\right)\ln\frac{W_1}{W_2} \tag{5.56}$$

where

$$\overline{V} = \frac{V}{V_{D_{\min}}}$$

$$\overline{V}_{w} = \frac{V_{w}}{V_{D_{\min}}}$$

The aircraft does not necessarily fly at E_m (where $\overline{V} = 1$) but at L/D consistent with the actual airspeed $\overline{V}$. E_m appears only as a consequence of introducing Eq. 3.41. The maximum range, in the presence of wind, is not obtained at E_m but at a different L/D, as indicated in Figure 5.9. In general, it is rather cumbersome and difficult to read off tangents from power, or thrust, curves. Rather, the tangent to the P_r curve can be related to the ground speed if a simple drag polar is available. In nondimensional form, and again by use of Eq. 3.36, the tangency condition is given by

$$\frac{d(P/W)}{d\overline{V}} = \frac{d}{d\overline{V}}\left[\frac{1}{2E_m}\left(\overline{V}^3 + \frac{1}{\overline{V}}\right)\right] = \frac{(P_r/W)}{\overline{V} \pm \overline{V}_w}$$

$$= \frac{1}{2E_m}\left(\overline{V}^3 + \frac{1}{\overline{V}}\right)\frac{1}{\overline{V} \pm \overline{V}_w} \tag{5.57}$$

Carrying out the differentiation and clearing the fractions, one finds

$$\frac{3\overline{V}^4 - 1}{\overline{V}} = \frac{\overline{V}^4 + 1}{\overline{V} \pm \overline{V}_w} \tag{5.58}$$

which gives the following inverse relationship:

$$\overline{V}_w = \frac{2\overline{V}\,(1 - \overline{V}^4)}{3\overline{V}^4 - 1} \tag{5.59}$$

It is inverse in the sense that a relationship of the form $\overline{V} = f(\overline{V}_w)$ is desired rather than $\overline{V}_w = f(\overline{V})$. However, Eq. 5.59 can be solved once and for all and is shown plotted on Figure 5.10 as the prop curve.

Thus, if the wind fraction $\pm\overline{V}_w$ is known, the best range speed $\overline{V}$ can be found from the curve and the range calculated from Eq. 5.56. A useful approximation for the solution of Eq. 5.59 is given by

$$3^{1/4}\overline{V} = \frac{4}{3} - \overline{V}_w\left(\frac{4}{9} - \frac{\overline{V}_w}{3}\right) \tag{5.60}$$

which can be used to explicitly evaluate the best range speed if $\overline{V}_w$ is known.

Jet Aircraft Similar to the developments leading to Eq. 5.57, the range for jet aircraft is given by

$$R = \frac{V_g}{C}\frac{L}{D}\ln\frac{W_1}{W_2} \tag{5.61}$$

where C_L = const and V = const. Using the same nondimensional parameters as for the propeller-driven aircraft, the range expression becomes

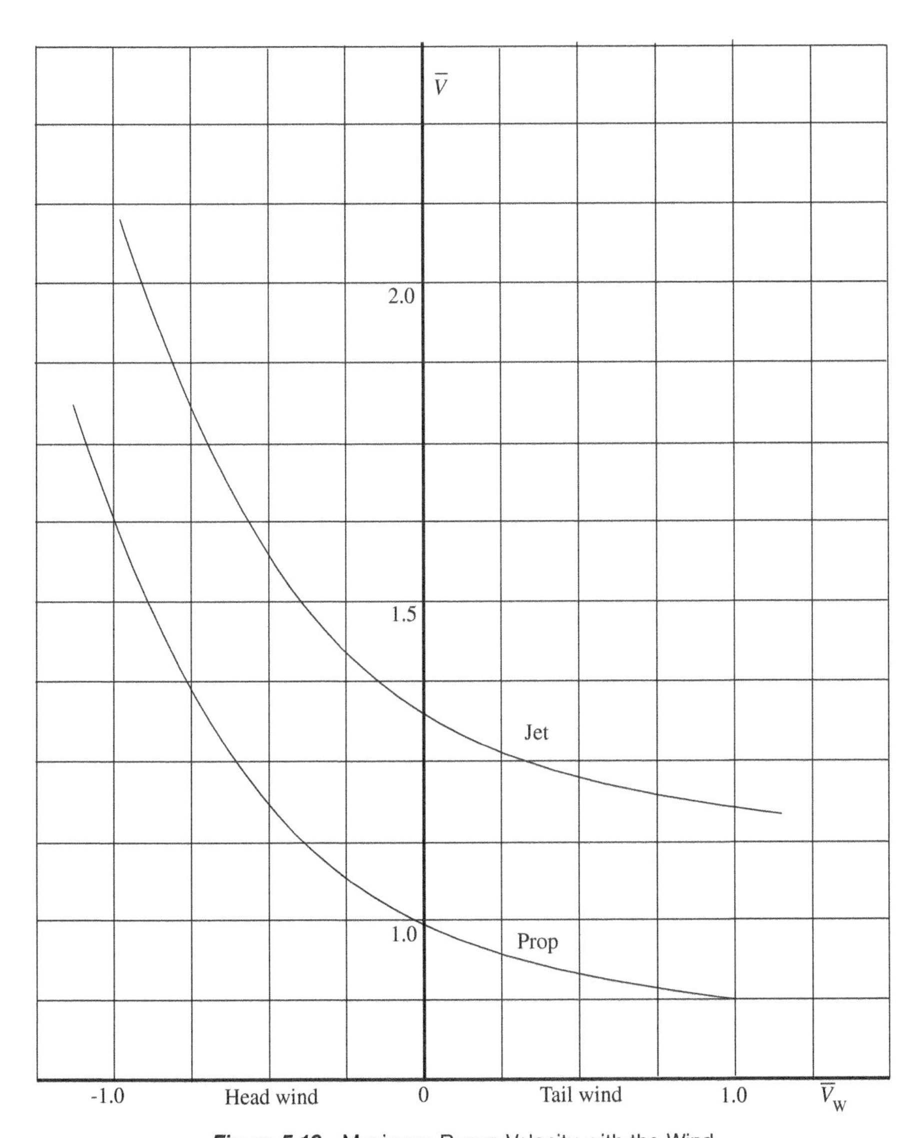

Figure 5.10 Maximum Range Velocity with the Wind

$$R = \frac{V_{D_{\min}}}{C}(\overline{V} \pm \overline{V}_w)\frac{L}{D}\ln\frac{W_1}{W_2}$$

$$= \frac{2V_{D_{\min}}}{C}\left(\frac{\overline{V} \pm \overline{V}_w}{\overline{V}^4 + 1}\right)\overline{V}^2 E_m \ln\frac{W_1}{W_2} \tag{5.62}$$

Again, the aircraft does not necessarily fly at E_m and $V_{D_{min}}$. These appear as a consequence of this particular nondimensional procedure. To find the airspeed that maximizes the range in the presence of wind, the tangency condition to the T_r curve needs to be established. Thus,

$$\frac{d(T_r/W)}{d\overline{V}} = \frac{d}{d\overline{V}}\left[\frac{1}{2E_m}\left(\overline{V}^2 + \frac{1}{\overline{V}^2}\right)\right] = \frac{(T_r/W)}{\overline{V} \pm \overline{V}_w}$$

$$= \frac{1}{2E_m}\left(\overline{V}^2 + \frac{1}{\overline{V}^2}\right)\left(\frac{1}{\overline{V} \pm \overline{V}_w}\right) \tag{5.63}$$

Upon simplification, one gets the following implicit equation for $\overline{V} = f(\overline{V}_w)$:

$$\overline{V}_w = \pm\frac{\overline{V}}{2}\left(\frac{\overline{V}^4 - 3}{1 - \overline{V}^4}\right) \tag{5.64}$$

This has been plotted on Figure 5.10 as the jet curve and can be given approximately as

$$\overline{V} = \frac{4}{3} - \overline{V}_w\left(\frac{4}{9} - \frac{\overline{V}_w}{3}\right) \tag{5.65}$$

The propeller and jet optimum flight velocity expressions, Eqs. 5.60 and 5.65, respectively, differ by the factor $3^{1/4}$ (see also Section 5.2.2), which makes these equations convenient to use. Thus, only one curve is sufficient in Figure 5.10.

EXAMPLE 5.5

Calculate the range and endurance of a propeller aircraft with 50 mph head wind at sea level. The aircraft has the following characteristics:

$$W_f = 15 \text{ percent of total weight}$$

$$\eta = 0.8$$

$$C = 0.5\ \frac{\text{lb}}{\text{HP-hr}}$$

$$W/S = 34$$
$$C_D = 0.022 + 0.0606 C_L^2$$

Calculate the range for C_L = const and V = const.

The range, from Eq. 5.56, can be written as

$$R = 375 \frac{\eta}{C} E_{\mathrm{m}} (\overline{V} - \overline{V}_{\mathrm{m}}) \left(\frac{2\overline{V}}{\overline{V}^4 + 1} \right) \ln \frac{W_1}{W_2}$$

$$= (\overline{V} - \overline{V}_{\mathrm{w}}) \left(\frac{2\overline{V}}{\overline{V}^4 + 1} \right) R_{\text{no wind}}$$

The no-wind range is determined at E_{m} and $V_{D_{\min}}$. Calculating these:

$$E_{\mathrm{m}} = \frac{1}{2\sqrt{kC_{D_0}}} = \frac{1}{2\sqrt{0.022 \times 0.0606}} = 13.7$$

$$V_{D_{\min}} = \sqrt{\frac{2W/S}{\rho}} \left(\frac{k}{C_{D_0}} \right)^{1/4} = \sqrt{\frac{2 \times 34}{0.002377}} \left(\frac{0.0606}{0.022} \right)^{1/4}$$

$$= 217 \frac{\text{ft}}{\text{sec}} = 148 \text{ mph}$$

provides

$$R_{\text{no wind}} = 375 \frac{\eta}{C} E_{\mathrm{m}} \ln \frac{W_1}{W_2} = 375 \frac{0.8}{0.5} 13.7 \ln \frac{1}{0.85} = 1{,}336 \text{ miles}$$

The wind fraction becomes $V_{\mathrm{w}} = 50/148 = 0.338$, which gives, for maximum range from Figure 5.10, $\overline{V} = 1.12$. The range, with the original no-wind range, can be calculated as follows:

$$R = (1.12 - 0.338) \left(\frac{2 \times 1.12}{1.12^4 + 1} \right) 1336 = 909.3 \text{ miles}$$

Thus, the original no-wind maximum range of 1,336 miles has been reduced by the 50 mph head wind to 909.3 miles. The flight takes place at

$$\overline{V} = 1.12$$

or

$$V = 1.12V_{D_{\min}} = 1.12 \times 148 = 166 \text{ mph}$$

and at $166 - 50 = 116$ mph ground speed. The actual flight lift-drag ratio is now

$$\frac{L}{D} = \frac{2\overline{V}^2}{\overline{V}^4 + 1} E_m = \frac{2 \times 1.12}{1.12^4 + 1} 13.7 = 11.9$$

As the flight speed is constant, the time to fly, or the endurance is

$$E = \frac{909.3}{116} = 7.83 \text{ hours}$$

As a first-order approximation that sometimes gives surprisingly good results, one can use

$$R_w = R_{\text{no wind}} \pm V_w E$$

For this case

$$R_w = 1336 - 50 \times \frac{1336}{148} = 1336 - 451 = 885 \text{ miles}$$

The preceding development and the example were carried out only for the cruise-climb flight schedule. Similar analysis applies for other flight paths (e.g., constant altitude, constant velocity cruise, as given by Eqs. 5.16 and 5.34). The constant ground speed needs to be replaced by its proper wind formulation, and Figure 5.10 can be used to find the correct flight (air) speed. As pointed out in Section 5.2, those rather cumbersome calculations are often replaced (or rather approximated) by the Breguet equations. Thus, that particular development will be omitted here.

5.6.2 Some Range and Endurance Comparisons

Maximum range and maximum endurance were calculated in the preceding sections for several different flight schedules. It is evident by inspection that these conditions do not occur simultaneously. For example, maximum range and maximum endurance for a jet aircraft (Breguet expressions) occur at $V_{D_{\min}}$ and at 1.316 $V_{D_{\min}}$, respectively. Moreover, jet aircraft endurance is independent of altitude. Thus, a jet flying at $V_{D_{\min}}$ at 20,000 ft and at sea level will have the same maximum endurance as $(L/D)_{\max}$ (with C being approximately constant). However, the plane flying at higher altitude will have more range due to higher airspeed. Similar arguments hold also for ranges flown at different schedules.

From Eq. 5.9, and assuming that the specific fuel consumption C is constant,

$$\frac{R_{\max}}{R_{E_{\max}}} = \frac{(L/D)_{\max}}{(L/D)_{E_{\max}}}$$

at $(L/D)_{\max}$

$$\frac{L}{D} = \frac{1}{2\sqrt{kC_{D0}}}$$

at $(L/D)_{E_{\max}}$

$$\frac{L}{D} = \frac{3}{4\sqrt{kC_{D0}}}$$

and

$$\frac{R_{\max}}{R_{E_{\max}}} = \frac{4\sqrt{kC_{D0}}}{2\sqrt{3kC_{D0}}} = \frac{2}{\sqrt{3}} = 1.155$$

TABLE 5.2 Comparison of Maximum Range and Endurance

	Prop	Jet
$\frac{E_{\max}}{E_{R_{\max}}}$	1.14	1.155
$\frac{R_{\max}}{R_{E_{\max}}}$	1.155	1.14

TABLE 5.3 Summary of Range and Endurance Equations

Prop			$V_{\max\ R,E}$	$C_{D0}\vert_{V_{\max}}$
Range (mi)	C_L = const	$375\,\frac{\eta_p}{C}\frac{L}{D}\ln\frac{W_1}{W_2}$	V_{E_m}	$C_{D_0} = kC_L^2$
	h = const, V = const	$750\,\frac{\eta_p E_m}{C}\left(\tan^{-1}[2kC_{L_1}E_m] - \tan^{-1}\left[2kC_{L_1}E_m\frac{W_2}{W_1}\right]\right)$	V_{E_m}	$C_{D_0} = kC_L^2$
Endurance (hr)		$1100\,\frac{\eta}{C}\sqrt{\frac{\rho SC_L}{2W_1}}\frac{L}{D}\left(\sqrt{\frac{W_1}{W_2}} - 1\right)$ $C = SFC$(lb/hr/HP)	$V_{P_{min}}$	$3C_{D_0} = kC_L^2$
Jet				
Range (mi)	C_L = const, V = const, cruise − climb	$\frac{V}{C}\frac{L}{D}\ln\frac{W_1}{W_2}$	$1.316V_{E_m}$	$C_{D_0} = 3kC_L^2$
	C_L = const, h = const	$\frac{2}{C}\sqrt{\frac{2W_1}{\rho SC_L}}\frac{L}{D}\left(1 - \sqrt{\frac{W_2}{W_1}}\right)$	$1.316V_{E_m}$	$C_{D_0} = 3kC_L^2$
	h = const, V = const	$\frac{2V\,E_m}{C}\left(\tan^{-1}[2kC_{L_1}E_m] - \tan^{-1}\left[2kC_{L_1}E_m\frac{W_2}{W_1}\right]\right)$	V_{E_m}	$C_{D_0} = kC_L^2$
Endurance (hr)		$\frac{1}{C}\frac{L}{D}\ln\frac{W_1}{W_2}$ $C = SFC$(lb/hr/T)	V_{E_m}	$C_{D_0} = kC_L^2$

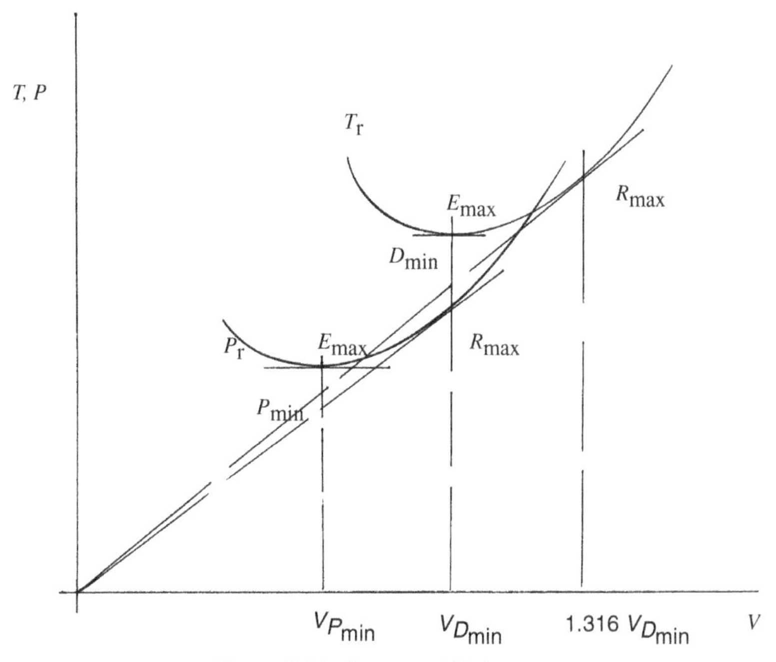

Figure 5.11 Range and Endurance

which states that a propeller aircraft maximum range is always 15.5 percent higher than the range flown at the best endurance condition. Similarly, a comparison of endurance for a propeller aircraft yields

$$\frac{E_{\max}}{E_{R_{\max}}} = \frac{C_L^{3/2}/C_D)_{\max}}{(C_L^{3/2}/C_D)_{(L/D)\max}}$$

In Section 5.5.1 it was shown that for maximum endurance

$$\frac{C_L^{3/2}}{C_D} = \frac{3^{3/4}}{4k^{3/4}C_{D0}^{1/4}}$$

Since, for $(L/D)_{\max}$, $C_L = \sqrt{C_{D0}/k}$, $C_D = 2C_{D0}$, then

$$\frac{E_{\max}}{E_{R_{\max}}} = \frac{3^{3/4}k^{3/4}2C_{D0}}{4k^{3/4}C_{D0}^{1/4}C_{D0}^{3/4}} = \frac{3^{3/4}}{2} = 1.14$$

The maximum endurance of a propeller airplane is 14 percent longer than the endurance at maximum range. Table 5.2 summarizes the results. Table 5.3 summarizes the general results for propeller and jet aircraft range and endurance. Figure 5.11 gives a graphical interpretation of the results found in Table 5.3.

PROBLEMS

5.1 Derive the jet aircraft range expression with the assumption that the cruise velocity is constant.

5.2 Calculate the range of the aircraft in Example 3.4 at 10,000 ft, assuming the specific fuel consumption of 0.5 lb/hr-HP.
Ans: 1,455 mi.

5.3 Calculate the maximum range of the aircraft in Example 3.3, assuming that it carries 2,000 lb fuel and is equipped with two J60 engines. What is its final flight altitude with 10 percent fuel remaining for descent and landing if the cruise climb schedule begins at 25,000 ft?
Ans: 797 and 723 mi.

5.4 Determine the ranges of the aircraft in Problem 5.3 with a 40 knot head wind and 40 knot tail wind.
Ans: 652 mi, head wind.

5.5 A jet aircraft has the following characteristics:

$$
\begin{aligned}
W &= 70{,}000 \text{ lb} \\
S &= 1500 \text{ ft}^2 \\
h &= 30{,}000 \text{ ft} \\
C_D &= 0.018 + 0.04\, C_L^2 \\
W_{\text{payload}} &= 9000 \text{ lb (dropped at destination)} \\
W_{\text{final}} &= 42{,}000 \text{ lb (including 5 percent fuel reserve)} \\
TSFC &= 0.6 \text{ (constant)}
\end{aligned}
$$

A long-range cruise drift-up technique is used. If an instantaneous altitude change takes place upon releasing the payload, calculate the distance at which the payload was dropped and the final altitude when reaching home base.

5.6 For the aircraft in Problem 5.5, calculate the distance at which the refueling should take place in order to increase the total range by 50 percent, 100 percent.

5.7 Show that

$$\frac{V}{T} = \frac{\sqrt{C_L}}{C_D}\sqrt{\frac{2}{\rho W\, S}}$$

which leads to Eq. 5.38 (see Section 5.2.2).

5.8 Boeing 707-320B, Intercontinental Series, weighs 320,000 lb at the beginning of its long-range cruise at 40,000 ft. Its drag polar, up to $M = .85$, is

$$C_D = .0185 + .036C_L^2$$

Other pertinent data:

$AR = 7.3$
$S = 2892\ \text{ft}^2$

Four JT3-1 engines

Calculate the amount of fuel needed to fly from London to New York, 3,500 mi.
Ans: about 88,000 lb.

5.9 Gates Learjet 56 Longhorn is a 13-place executive transport with two 3,700 lb turbofan engines. It has a maximum cruise weight of 20,000 lb, $S = 265\ \text{ft}^2$, $AR = 7.2$, e = .85, $C_{D_0} = .018$, cruise airspeed 508 mph (.77M). With a normal fuel load of 5,500 lb, it has a range of 3,000 mi (sfc = .75), more than sufficient capacity to fly from Los Angeles to New York (2,450 mi). During the last flight, the pilot, planning to capitalize on a favorable jet stream, took on an extra 1,500 lb of commercial cargo and only 4,000 lb of fuel, which normally would be insufficient to reach New York without an appreciable tail wind. Determine the following:

a. What would be the minimum jet stream speed that the pilot needed to make good the distance?

b. At what airspeed would the pilot fly to get the best range?
Ans: a. 77 mph, b. 486 mph.

5.10 A jet aircraft weighs 29,950 lb at the beginning of a constant altitude cruise at 20,000 ft altitude. $S = 640$ ft^2, $sfc = .87$. From the thrust setting it can be estimated that the thrust is initially 2,650 lb. Calculate the range if the cruise fuel weighs 8,150 lb. The drag polar is

$$C_D = .0255 + .0616C_L^2$$

Ans: 1117 mi.

5.11 An A-4 aircraft is being tested at 5,000 ft altitude. The aircraft is equipped with one J-52 engine capable of delivering a thrust of 8,500 lb up to $M = .8$. Fuel flow rate is given in terms of the thrust and Mach number by $\dot{w}_f = .9T(1+.36M)$ lb/hr. Aspect ratio is 2.91, $S = 260$ ft^2. Weight of the aircraft, at the beginning of the flight, is 16,000 lb, and the available fuel weighs 1,700 lb. $C_D = .022 + .19C_L^2$. The ground radar measures the speed to be $.6M$. Assuming that the simple drag polar holds, estimate the aircraft range under these (constant) flight conditions.
Ans: about 240 mi.

5.12 A four-engine turbojet has the following characteristics: $C_{D_0} = .018$, $M_{cr} = .77$, $S = 1{,}500$ ft^2, $AR = 5.5$, $e = .82$, $W/S|_{TO} = 70$ $C_{L_{TO}} = 2.0$, $T = 1{,}600\ \sigma$ lb/engine, C = .87 lb$_m$/lb$_f$/hr.

The aircraft uses 3,000 lb fuel to take off and to climb to 30,000 ft. Determine:

a. R_{max} for constant altitude flight at 30,000 ft for 24,000 lb available cruise fuel

b. The velocity at the destination

5.13 The following information is known about 727-200:

$$
\begin{aligned}
W &= 135{,}000 \text{ lb}\\
S &= 1{,}700 \text{ ft}^2\\
AR &= 6.9\\
e &= .83\\
T &= 15000\sigma^{.8}/\text{engine, 3 engines}\\
TSFC &= .95 \text{ lb}_m/\text{hr}/\text{lb}_T\\
C_{D_0} &= .019,\ C_{D_0\,\text{land}} = .092 \text{ (including air brakes)}\\
C_{L_{max}} &= .96,\ C_{L_{max\ flaps}} = 2.6,\ C_{LTO} = 2.1
\end{aligned}
$$

Determine:

a. How much fuel is needed to fly from Baltimore to Chicago (700 miles) using the maximum range schedule at constant altitude

b. Upon arriving in Chicago, how much thrust is required to maintain a glide slope of 2 degrees
Ans: 13,265 lb, 3739 lb.

6

Nonsteady Flight in the Vertical Plane

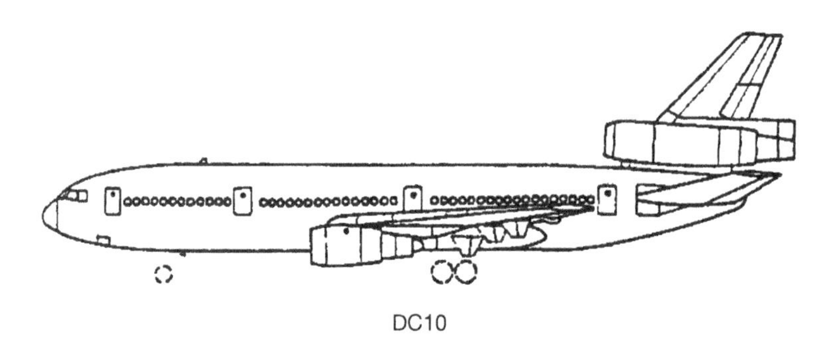

DC10

6.1 TAKE-OFF AND LANDING

The quasi-steady analysis presented in previous chapters determines the general bounds of the aircraft performance: stall, maximum velocity, ceiling, climb, range, and endurance. Although corrections were included for speed variations, the average steady-state nature of the phenomena permitted solutions without recourse to path-dependent integration and without the need for direct consideration of accelerations. Take-off and landing are accelerated operations involving velocity variations from zero to safe flight speed, thus requiring a more exacting analysis of equations of motions. Although take-off and landing represent only a small portion of the total operation of an aircraft, performance of these two phases is considered very important due to two entirely different reasons. First, a great majority of accidents (mostly attributed to pilot error) occur during landing or take-off. Second, it is the take-off portion that establishes the engine sizing (in conjunction with air worthiness requirements) for design of civil aircraft. For multiengine design, the critical condition is provided by the failure of one or more engines at the least favorable moment.

Thus, owing to the variability of the pilot techniques human element, it is not practical to introduce too many exacting mathematical complications in the analysis of take-off or landing performance. On one hand, experience indicates that it is practical to introduce those simplifying assumptions without destroying the fundamental aspects of the phenomena, which lead to analytical solutions permitting inclusion of all important factors affecting the performance. On the other hand, design studies may require that extensive computer calculations be carried out to provide systematic studies concerning the effect of various parameters on aircraft performance. The wealth of resulting numerical detail can lead to useful design results, but there is also a real danger of obscuring the basic features of the problem. To support the detailed numerical studies an analysis is needed that is sufficiently simple yet permits a good overview of all pertinent features of the problem; this is the purpose of the current chapter.

6.2 TAKE-OFF ANALYSIS

The take-off analysis can be conveniently divided into two phases: the ground run and the air phase. The air phase begins when the aircraft becomes airborne and lasts until the aircraft has reached a safe flight condition. The safe flight condition is defined as the condition where the aircraft has cleared a specified height and commences to climb at a steady rate. This height is called obstacle height or screen height—FAA specifies 50 ft. For more details, see FAR23 and FAR25 series or applicable MILSPEC's.

Depending on the type of aircraft and the take-off performance capabilities, the airborne phase may be divided into two parts: a transition phase and a steady climb phase. If the 50-ft height is reached before steady flight is achieved, the airborne phase is approached as one continuous operation. Due to operational safety aspects FAA uses also the concept of balance field length (BFL). At its simplest, it can be defined as a total length when the distance required for take-off is equal to a distance needed to come to a complete stop on the ground (aborting the take-off). Its real intent is to assure safe landing or take-off distances for multiengined aircraft experiencing engine(s) failure at take-off. Very little generalization is possible here. In practice, the large number of possible cases arising from different aircraft subject to different FAR rules at a variety of flight conditions require highly individual and detailed treatment. Although the necessary methodology is displayed below, it is not practical to pursue these details within the intent and given scope of the topic treatment here.

The total take-off distance is calculated as the ground distance to clear the screen height as shown in Figure 6.1, and is the sum of the following four distances:

s_g—ground run
s_r—rotation distance
s_t—transition distance
s_c—climb distance to reach screen height

In the following sections these distances will be evaluated separately, as they represent individual portions of the take-off operation. However, they are each subject to some of the following factors:

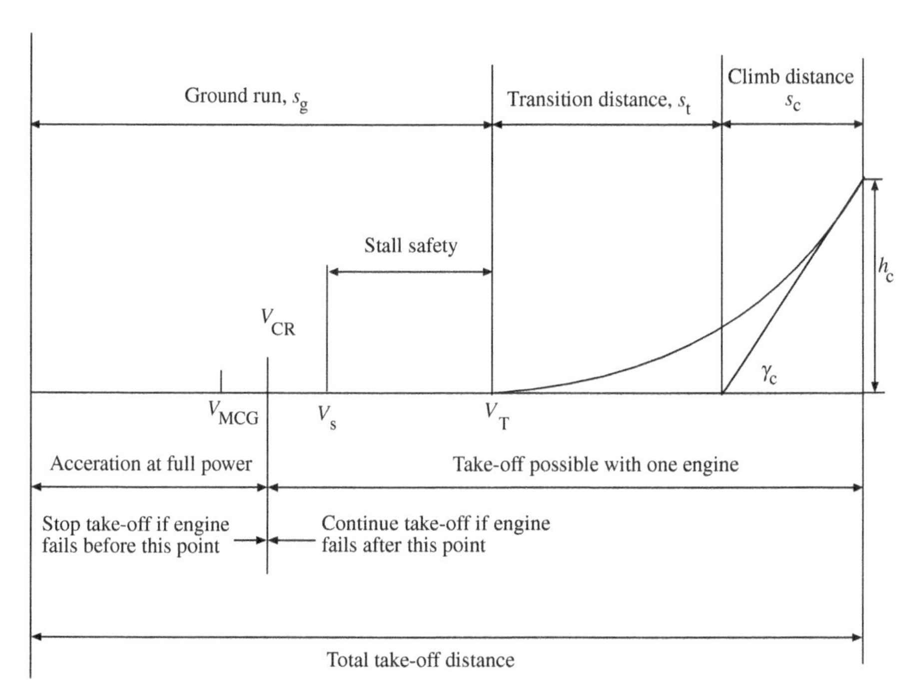

Figure 6.1 Take-off Distances

β—runway slope or gradient
V_w—wind speed, engine failure during take-off

These factors will be considered inasmuch as practical results can be obtained.

6.2.1 Ground Run

Ground roll can be defined as the distance between the points where the brakes are released and where the take-off velocity, V_T is reached. During the ground roll the velocity increases progressively through a number of (operationally) important velocities, as shown in Figure 6.1, and they are stated in the order in which they occur on a multiengine aircraft.

1. V_{MCG}, minimum control speed on the ground. If an engine failure occurs above this speed the pilot is able to maintain control and a straight path in the original direction without retrim or reduction of thrust.
2. V_{CR}, critical speed. One engine failure at this speed permits the pilot to continue take-off to screen height or stop, in the same distance. Engine failure at lower speed requires stopping the aircraft, while a failure at higher speed means that the take-off be continued.
3. V_s, stall speed (decision speed). This is calculated in the take-off configuration.
4. V_T, take-off speed. After leaving the ground, the airspeed should not be allowed to fall below this speed. Also called take-off safety speed, it is usually chosen as $1.2 V_s. The aircraft can be lifted off the ground (unstick, or getaway point) at any speed above V_s, but no appreciable height can be gained until V_T has been reached. It is questionable whether completing ground roll with wheels off the ground will shorten the ground phase at all since, while in ground contact, above V_s there will be very little load on the wheels anyway and the induced drag is reduced due to favorable ground effect. Lifting the aircraft out of the ground effect early may increase the drag, and thus also increase the distance in which V_T is reached.

While an aircraft is in airborne motion, the governing equations are those given by Eqs. 2.9 through 2.13. Two additional forces appear in the ground phase: the reaction force R of the runway on the aircraft and the tangential force due to rolling friction μR; see Figure 6.2.

Average values for the friction coefficient μ are given in Table 6.1. They can be used for either take-off or landing. Thus, if it is assumed that the runway is horizontal ($\gamma = 0$) and that the weight variation due to fuel consumption can be neglected, the equations of motion can be rewritten as

$$\dot{x} = V \tag{6.1}$$

$$\frac{\dot{V}W}{g} = T - D - \mu R \tag{6.2}$$

$$0 = L - W + R \tag{6.3}$$

If the reaction R is now eliminated and the velocity V is selected as a new independent variable, one obtains the following differential equations for the distance and time ($s \equiv x$):

$$\frac{ds}{dV} = \frac{VW/g}{T - D - \mu(W - L)} \tag{6.4}$$

$$\frac{dt}{dV} = \frac{W/g}{T - D - \mu(W - L)} \tag{6.5}$$

In general, since T, D, and L are functions of velocity, exact solutions are possible only by carrying out numerical integration.

Several schemes are available in literature for integrating these expressions. The two most common assumptions have made use of either

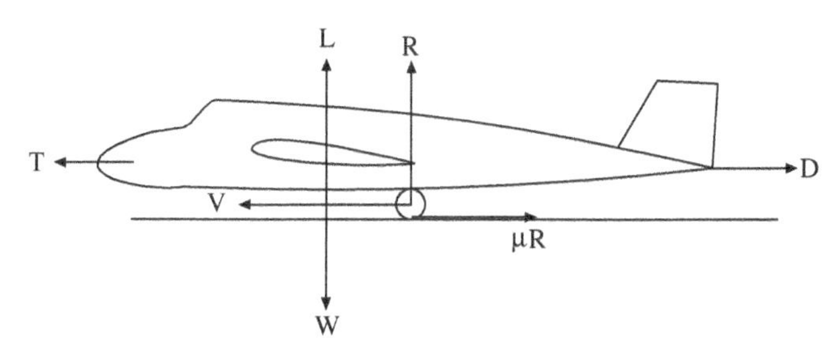

Figure 6.2 Forces on an Aircraft During Take-off

TABLE 6.1 Average Coefficient of Friction Values

Type of Surface	Average Coefficient No Brakes	Average Coefficient Full Brakes
Concrete	0.02 to 0.05	0.4 to 0.6
Hard Turf	0.05	0.4
Soft Turf	0.07	
Wet Concrete	0.05	0.3
Wet Grass	0.10	0.2
Snow/Ice Covered Field	0.02	0.7 to 1.0

a constant average force idea (the denominator is the acceleration force) or an average acceleration concept. Although such methods require additional assumptions concerning individual parameters, they do provide practical engineering approximations. However, this results in hiding the trends and basic features of the problem, thus preventing a good overview of the role of the significant parameters involved.

A fully equivalent and more flexible solution becomes available if one assumes that C_L and C_D are constants during the ground phase and the take-off portion of the thrust can be approximated (by simple curve-fitting procedure) by an equation of the form

$$T = T_o + BV + CV^2 \tag{6.6}$$

Then the expressions for distance and time become

$$ds = \frac{V\,dV}{aV^2 + bV + c} \tag{6.7}$$

$$dt = \frac{dV}{aV^2 + bV + c} \tag{6.8}$$

where

$$a = -\frac{\rho g S}{2W}(C_D - \mu C_L) + \frac{Cg}{W}$$

$$b = \frac{Bg}{W}$$

$$c = g\left(\frac{T_o}{W} - \mu\right)$$

Since the solutions reappear in landing calculations and due to the need to evaluate the integral between any two speeds as the thrust may vary substantially when the take-off assisting rockets are fired or an engine fails, the general solutions are given by (see any good table of integrals) between two arbitrary limits V_1 and V_2 as

$$s_g = \frac{1}{2a}\ln\left(\frac{aV_2^2 + bV_2 + c}{aV_1^2 + bV_1 + c}\right) + \frac{b/a}{\sqrt{b^2 - 4ac}}\left(\tanh^{-1}\frac{2aV_2 + b}{\sqrt{b^2 - 4ac}} - \tanh^{-1}\frac{2aV_1 + b}{\sqrt{b^2 - 4ac}}\right) \tag{6.9}$$

or

$$s_g = \frac{1}{2a}\ln\frac{aV_2^2 + bV_2 + c}{aV_1^2 + bV_1 + c} + \frac{b/2a}{\sqrt{b^2 - 4ac}}\ln\left(\frac{(1 + a_2)(1 - a_1)}{(1 - a_2)(1 + a_1)}\right) \tag{6.10}$$

where

$$a_2 = \frac{(2aV_2 + b)}{\sqrt{b^2 - 4ac}}$$
$$a_1 = \frac{(2aV_1 + b)}{\sqrt{b^2 - 4ac}}$$

The time for take-off becomes

$$t = \frac{2}{\sqrt{b^2 - 4ac}}\left(\tanh^{-1}\frac{2aV_1 + b}{\sqrt{b^2 - 4ac}} - \tanh^{-1}\frac{2aV_2 + b}{\sqrt{b^2 - 4ac}}\right) \tag{6.11}$$

$$= \frac{1}{\sqrt{b^2 - 4ac}}\ln\left(\frac{(1 + a_1)(1 - a_2)}{(1 - a_1)(1 + a_2)}\right) \tag{6.12}$$

Simple calculations will show that, in most cases, the quantity $(b^2 - 4ac)$ is positive. Thus, the choice of inverse hyperbolic tangent solution. In case $4ac > b^2$ (this may happen if large reverse thrust is applied, $T_o < O$). Then the inverse hyperbolic tangent must be replaced by inverse tangent function, and the radical in the denominators will be replaced by $\sqrt{4ac - b^2}$. Furthermore, Eqs. 6.10 and 6.12 cease to be valid for this case, as they represent the equivalent functions *only* for the case of inverse hyperbolic tangent function.

To bring out the factors that are essential for establishing the ground run distance, consider the following (rather general) case: normal unassisted, level runway, no wind, starting at rest, constant thrust take-off. Thus (see Eq. 6.6):

$$V_1 = 0 \qquad B = 0$$
$$T = T_o \qquad C = 0$$

and one obtains from either Eqs. 6.11 or 6.12

$$s_g = \frac{1}{2a} \ln \frac{aV_2^2 + c}{c} = \frac{W/S}{\rho g\,(C_D - \mu C_L)} \ln \left[\frac{1}{1 - \dfrac{C_D - \mu C_L}{(T_o/W - \mu)C_{L_T}}} \right] \tag{6.13}$$

The take-off lift coefficient C_{L_T} is obtained from the condition of the take-off speed when the aircraft just leaves the ground:

$$C_{L_T} = \frac{2W/S}{\rho V_T^2}, \quad V_T \equiv V_2 \tag{6.14}$$

The coefficients C_D and C_L are average values evaluated during the ground run. Eq. 6.13 can be further simplified by expanding the logarithm as

$$\ln x = \frac{x-1}{x} + \frac{1}{2}\left(\frac{x-1}{x}\right)^2 + \cdots \qquad x > \frac{1}{2} \tag{6.15}$$

Letting now $x = 1/(1 - Z)$, with Z being the term in the square brackets in Eq. 6.13, then

$$\ln \frac{1}{1 - Z} = Z + \frac{Z}{2} + \cdots \tag{6.16}$$

and Eq. 6.13 can be written as

$$s_g = \frac{W/S}{\rho g \left(\dfrac{T_o C_{L_T}}{W} - \mu\right)} \left[1 + \frac{1}{2} \frac{C_D - \mu C_L}{\left(\dfrac{T_o}{W} - \mu\right) C_{L_T}} + \cdots \right] \tag{6.17}$$

where the significant remaining term in the brackets is usually less than 0.1, and the discarded terms are entirely negligible.

NOTE

For rapid calculations and comparative estimates, Eq. 6.17 is simplified to read

$$s_g = \frac{W/S}{\rho g C_{L_T} (T_o/W)}$$

$$= \frac{1.44(W/S)}{\rho g C_{L_m} (T_o/W)} \tag{6.18}$$

Eq. 6.18 brings out all the salient features of the take-off ground run. It shows that the ground run is directly proportional to the aircraft wing loading W/S and the altitude of the runway. It varies inversely with the take-off lift coefficient and T_o/W (μ being about 0.02, and thus may be ignored). Increase in ambient temperature works through density ($\rho \propto 1/T$) and through degradation of engine thrust to increase the take-off distance.

Calculation of the take-off ground run is not very sensitive to aircraft attitude and, thus, to evaluation of C_L, especially since it appears in combination with rolling friction coefficient μ. However, an optimum lift coefficient can be determined that gives theoretically a minimum ground distance and provides some guidance for selection of the ground run lift coefficient. When differentiating Eq. 6.13 with respect to C_L, it leads to the requirement that

$$\frac{da}{dC_L} = \frac{d}{dC_L}(C_{D_0} + kC_L^2 - \mu C_L) = 0 \tag{6.19}$$

and

$$2kC_L - \mu = 0 \tag{6.20}$$

It then follows that

$$C_{L_g} = \frac{\mu}{2k} = \frac{\mu \pi AR e}{2} \tag{6.21}$$

for the ground run lift coefficient that tends to give the shortest distance.

The drag coefficient is evaluated by means of Eq. 6.21, and C_{D0} should include the additional drag due to landing gear, flaps, and the ground effect. The effect of ground, or water, is to reduce the aircraft downwash and the strength of tip vortices that makes the aircraft behave as if its aspect ratio were increased. The spanwise lift distribution is then altered in a favorable manner to produce more lift and less induced drag and a smaller angle of attack is required to produce the same lift, which is also significant during the landing operations. Thus, to account for the ground run, the induced drag may be reduced by a factor of 0.6:

$$C_D = C_{D0} + 0.6kC_L^2$$

In transition and climb-out phases, the ground effect is generally ignored.

For approximate preliminary design estimates the following expression can be obtained from Eq. 6.18 by ignoring the terms in the brackets and the rolling coefficient in comparison with T/W, and by introducing the take-off velocity:

$$V_T^2 = \frac{2W}{\rho S C_{L_T}} \tag{6.22}$$

as

$$s_g = \frac{W V_T^2}{2 g T_o} \tag{6.23}$$

where W is the take-off weight. For use with Eq. 6.22, the thrust is evaluated at $(0.7 - 0.8)\ V_T$.

Summarizing: Eqs. 6.23, 6.18, 6.17, 6.13, and 6.9 calculate the ground run in an increasing order of accuracy. Eqs. 6.23 and 6.22 are practically the same first-order approximations. Eq. 6.13 marginally improves the accuracy over the simplified version of Eq. 6.17.

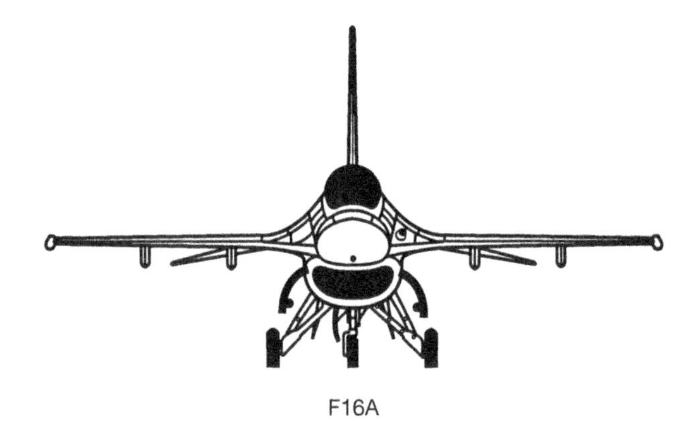

F16A

6.2.2 Rotation Distance

At the end of ground roll and just prior to going into transition phase, most aircraft are rotated to achieve an angle of attack to obtain the desired take-off lift coefficient C_L. Since the rotation consumes a finite amount of time (1–4 seconds), the distance traveled during rotation s_r, must be accounted for by using

$$s_r = V_T \Delta t \tag{6.24}$$

where Δt is usually taken as 3 seconds.

EXAMPLE 6.1

A twin-engine jet aircraft has the following characteristics at sea level:

$$W = 50{,}000 \text{ lb}$$
$$T = 5000 - 3.28\text{V per engine}$$
$$S = 1500 \text{ ft}^2$$
$$C_{L_{\max}} = 1.5$$
$$C_D = 0.02 + 0.05C_L^2$$

Calculate the take-off distance on a smooth, dry concrete runway (μ = .02) at sea level for no-wind conditions.

Ground run is calculated from Eq. 6.9. Assuming $V_T = 1.2V_s$, one obtains

$$C_{L_T} = \frac{C_{L\max}}{1.44} = \frac{1.5}{1.44} = 1.04$$

and

$$V_T^2 = 1.44 \frac{2W}{\rho S C_{L_{max}}} = 1.44 \frac{2 \times 50{,}000}{0.002377 \times 1500 \times 1.5} = 26{,}925 \frac{\text{ft}^2}{\text{sec}^2}$$

Ground run lift coefficient is obtained from Eq. 6.21:

$$C_L = \frac{\mu}{2k} = \frac{0.02}{2 \times 0.05} = 0.2$$

which gives for the drag coefficient:

$$C_D = 0.02 + 0.05(0.2)^2 = 0.022$$

For evaluating the ground run from Eq. 6.20 the coefficients a, b, and c are

$$a = \frac{-\rho g S}{2W}(C_D - \mu C_L) = \frac{-0.002377 \times 32.2 \times 1{,}500}{2 \times 50{,}000} \times (0.022 - 0.02 \times 0.2) = -2.02 \times 10^{-5}$$

$$b = \frac{Bg}{W} = -\frac{6.56 \times 32.2}{50{,}000} = -4.225 \times 10^{-3}$$

where $B = 2 \times 3.28$

$$c = g\left(\frac{T_o}{W} - \mu\right) = 32.2\left(\frac{10{,}000}{50{,}000} - 0.02\right) = 5.796$$

The ground run is then from Eq. 6.9, $V_1 = 0$, $\sqrt{b^2 - 4ac} = 0.022$

$$s = \frac{-10^5}{2.02 \times 2} \ln \left(\frac{-2.02 \times 10^{-5} \times (164)^2 - 4.225 \times 10^{-3} \times (164) + 5.796}{5.796}\right) + \frac{-4.225 \times 10^{-3}}{-2.02 \times 10^{-5} \times 0.022} \times \left[\tanh^{-1} \frac{2 \times (-2.02 \times 10^{-5}) \times 164 - 4.225 \times 10^{-3}}{0.022} - \tanh^{-1} \frac{-4.225 \times 10^{-3}}{0.022}\right] = 5{,}938 - 3{,}288 = 2{,}650 \text{ ft}$$

Two comparisons are now of great interest. First, if it is assumed, as it is often the case, that the take-off thrust is constant ($B = 0$, $b = 0$), then the second term in Eq. 6.9 drops out and the ground run becomes

$$s = \frac{-10^5}{2.02 \times 2} \ln \frac{-2.02 \times 10^{-5} \times (164)^2 + 5.796}{5.796} = 2{,}436 \text{ ft}$$

Second, using the approximate result, Eq. 6.16, one gets

$$s = \frac{1}{\dfrac{\rho g S}{W} C_{L_T} \dfrac{T_o - \mu}{W}} \left(1 + \frac{1}{2}\left[\frac{C_D - \mu C_L}{\left(\dfrac{T_o}{W} - \mu\right) C_{L_T}}\right]\right)$$

$$= \frac{1}{0.002296 \times 1.04 \times 0.18}\left(1 + \frac{0.5 \times 0.0176}{0.18 \times 1.04}\right)$$

$$= 2{,}326 \times 1.04 = 2{,}435 \text{ ft}$$

Rotation distance is evaluated from Eq. 6.23:

$$s_r = 3V_T = 3 \times 164 = 492 \text{ ft}$$

Thus, the total distance on the ground is $s_g = 2{,}650 + 492 = 3{,}142$ ft.

6.2.3 Transition Distance*

Evaluation of the transition distance is cumbersome and results tend to be unreliable due to various factors like landing gear retraction, varying ground effect, different piloting techniques, and so on. In addition, the number of parameters tends to be large or parameters appear in the analysis that can be evaluated only by experimental means. It becomes significant only for modestly powered and/or extremely loaded aircraft. Thus, a method that is convenient and relatively simple to apply with acceptable accuracy is obtained from Eq. 6.8 by recognizing that $\mu = 0$ for the airborne phase, and by assuming that $T - D =$ constant. Thus, by direct integration of Eq. 6.8 one gets

$$s_t = \frac{W}{2g}\frac{V_a^2 - V_T^2}{T - D} \tag{6.25}$$

where V_a is the airspeed at the end of transition, and T and D are to be evaluated at the average velocity $(V_a + V_T)/2$ during transition. The inherent drawback of this method is the fact that the airborne velocity V_a, at the end of transition is not known and must be either assumed or iteratively determined. Furthermore, the process is complicated by the fact that the transition distance is not the distance in which the steady-state climb path is achieved. Rather, it is the distance from the lift-off point to the location where the climb path slope intersects the ground (Figure 6.1). However, it should be recognized that high-performance aircraft with high thrust-weight ratios will reach the screen height before transition is actually completed.

The following steps can be used to facilitate calculation procedure for determining s_t:

1. $V_T = 1.2V_s$ is assumed.
2. $V_a = (1.2 + d)V_s$ is chosen, where d may be taken anywhere between 0.05 to 0.15.
3. Evaluate T and D at $(V_a + V_T)/2$.
4. Calculate s_t from Eq. 6.25.
5. Calculate the average transition γ_t from

$$\sin \gamma_t = \left.\frac{T - D}{W}\right|_{\text{jet}} \tag{6.26}$$

$$\sin \gamma_t = \left.\frac{P_a - DV_{avg}}{WV_{avg}}\right|_{\text{prop}} \tag{6.27}$$

6. Check the height h_t reached when achieving the (assumed) airspeed V_a from

$$h_t = s_t \tan \gamma \tag{6.28}$$

Then:

a. If $h_t > h_{screen}$, the screen height has been passed during the transition and s_t calculated in step 4 above is too large. Thus, a new and smaller V_a must be assumed and the procedure repeated until $h_t \simeq h_{screen}$.

b. If $h_t < h_{screen}$, it is assumed that the transition is completed and the climb distance can be calculated for the resulting steady climb phase to the screen height.

The climb distance is evaluated from the following (see Figure 6.1):

$$s_c = \frac{h_c}{\tan \gamma_c} \simeq \frac{h_c}{\gamma_c} \tag{6.29}$$

For small angles of climb, with $L = W$, γ_c can be written

$$\gamma_c = \frac{T}{W} - \frac{D}{L} = \frac{T}{W} - \frac{C_{D0}}{C_{L_c}} - kC_{L_c} \tag{6.30}$$

whence

$$s_c = \frac{h_c}{T/W - C_{D0}/C_{L_c} - kC_{L_c}} \tag{6.31}$$

Strictly speaking, the climb-lift coefficient should be evaluated at a climb speed higher than V_a since the aircraft will continuously pick up speed. For practical purposes, however, evaluating C_{L_c} at V_a will provide satisfactory results.

Thus, summarizing, the basic take-off distance can be calculated from Eqs. 6.11, 6.24, 6.25, and 6.31, or

$$S_{T.O.} = s_g + s_r + s_t + s_c \tag{6.32}$$

EXAMPLE 6.2

For Example 6.1, determine the transition distance and the total take-off distance.

Transition distance is calculated according to procedures given in Section 6.2.2.

1. $V_T = 1.2V_s = 1.2 \times 136.7 = 164$ ft/sec
2. Assume $d = 0.05$, then $V_a = 1.25V_s = 1.25 \times 136.7 = 170.9$ ft/sec

3.

$$\frac{V_a + V_T}{2} = 167.5 \text{ ft/sec}$$

$$C_L = \frac{C_{L\max}}{(1.25)^2} = \frac{1.5}{(1.25)^2} = 0.96$$

$$C_D = 0.02 + 0.05(0.96)^2 = 0.066$$

$$D = 1/2(0.002377)(167.5)^2(1500)(0.066) = 3{,}301 \text{ lb}$$

$$T = 2(5{,}000 - 3.28(167.5)) = 8901 \text{ lb}$$

4.

$$s_t = \frac{W}{2g}\frac{V_a^2 - V_T^2}{T - D} = \frac{50{,}000}{2 \times 32.2}\frac{170.9^2 - 164^2}{8{,}901 - 3{,}301} = 320 \text{ ft}$$

5.

$$\sin\gamma = \frac{T - D}{W} = \frac{8{,}901 - 3{,}301}{50{,}000} = 0.112$$

$$\gamma = 6.4°$$

6. $h_t = s_t \tan\gamma = 320 \times \tan 6.4° = 36$ ft, which is less than the screen height of 50 ft; thus, one can proceed to calculation of the climb phase distance.

The climb distance is evaluated as follows:

$$s_c = \frac{h_s}{\gamma_c}$$

$$\gamma_c = \frac{T}{W} - \frac{D}{L} = \frac{8{,}901}{50{,}000} - \frac{0.066}{0.96} = 0.1093$$

$$s_c = \frac{50}{0.1093} = 458 \text{ ft}$$

Finally, then the total take-off distance is

$$s_{T.O.} = s_g + s_r + s_t + s_c = 2650 + 492 + 320 + 458 = 3{,}920 \text{ ft}$$

6.2.4 Take-off Time*

The time to take off consists of the same individual segments as shown for the ground run in Eq. 6.32. The duration of the ground run follows directly from Eq. 6.11 with $V_1 = 0$ as

$$t = \frac{2}{\sqrt{b^2 - 4ac}} \left(\tanh^{-1} \frac{b}{\sqrt{b^2 - 4ac}} - \tanh^{-1} \frac{2aV_T + b}{\sqrt{b^2 - 4ac}} \right) \tag{6.33}$$

with the constants a, b, c given with Eq. 6.6. If the thrust is constant, then $b = 0$, and the above can be simplified to give

$$t_g = \frac{V_2}{g(T_o/W - \mu)} \tag{6.34}$$

For the rotation it is assumed that $t_r =$ const. Calculation of the transition phase starts with Eq. (6-2) as now, $\mu = 0$. Then, assuming that $T - D \equiv$ const,

$$t_t = \frac{W}{g} \int_{V_T}^{V_a} \frac{dV}{T - D} = \frac{W}{g} \frac{V_a - V_T}{T - D} \tag{6.35}$$

The airspeed V_a can be calculated by the procedure given in the last section.

The time during the climb-out phase is given by

$$t_c = \frac{s_c}{V_a} \tag{6.36}$$

where s_c is given by Eq. 6.31. The total time to reach height is obtained from

$$t = t_g + t_r + t_t + t_c \tag{6.37}$$

6.2.5 Factors Influencing the Take-off

For the take-off portion, it remains now to evaluate the various effects that influence the take-off distance.

Runway Gradient The effect of runway gradient, β, is to modify the effective thrust in Eqs. 6.7 and 6.8 by a factor of βW. Thus, the coefficient c in Eq. 6.7 becomes

$$c = 2g\left(\frac{T_o}{W} - \mu \pm \beta\right) \tag{6.38}$$

where + and − are used for downhill or uphill slope, respectively.

Engine Failure In multiengine design, one of the critical cases occurs when an engine fails at take-off. Then it is needed to know the distance to complete take-off run to the screen height, to bring the aircraft to a stop, or to evaluate V_{CR} for a given runway length.

The effect of engine failure can be accounted for by derating the thrust-weight ratio by a factor f, which stands for the fraction of thrust available after one (or) more engine failures. Thus, wherever T/W or T_o/W appears it is to be replaced by $f(T/W)$ after the engine failure.

Although an engine failure may occur anywhere during the take-off, for simplicity it is assumed that the failure will occur during the ground run up to the lift-off at V_T. This will require dividing the take-off portion into two sections and applying Eq. 6.11:

a. $V_1 = 0$, $V_2 = V_f$, during the starting portion, where V_f is the speed at which engine failure occurs.
b. $V_1 = V_f$, $V_2 = V_T$, for the section up to the lift off if the take-off is to be continued. If take-off is aborted then the procedures for landing run apply (see next section). For this portion of take-off T and D will be evaluated at an average velocity $V_m = (V_f + V_T)/2$.

This concept can be extended to cover engine failure in any take-off phase.

Thrust Augmentation There are several ways for shortening the ground run or achieving satisfactory take-off performance under unfavorable conditions. Ground run can be reduced by lifting off the runway at a lower speed since the ground distance is directly proportional to V_T^2 (note the group $W/\rho SC_{L_T}$ in Eq. 6.23). V_T can be reduced if C_{L_T} can be increased, by means of flaps, slats, boundary layer control, and so on. A point of diminishing return is reached quickly, however, since flap deflection causes drag to increase, which may, in turn, increase the take-off distance. Thus, a careful choice and use of flaps deflection is necessary during the take-off.

Thrust augmentation by various means (water injection to reduce effective inlet temperature, after-burner, auxiliary jets, or rockets) is very effective in reducing the ground run, as it is inversely proportional to thrust; see Eq. 6.18. If thrust augmentation is used throughout the entire ground run, the calculations proceed in the same manner as in the preceding sections, with the thrust given at the new augmented level. Often the augmentation time is shorter than the ground run (especially in case of rocket firing) and the calculations must be carried out in several steps. Moreover, it is necessary then to determine when the augmenting unit should be fired.

The augmenting units usually produce constant thrust, which is applied at a varying speed. It is most effective when the power produced, or work done, is maximum. The available work is

$$W = \Delta T \times V \times \Delta t \tag{6.39}$$

and since ΔT and Δt are fixed by the engine characteristics, the work done increases with increasing velocity. Clearly, then, the thrust augmentation should be used during the latter part of ground run with the burnout to coincide with the lift-off.

Assuming that the unit is fired at the time a ground speed of V_1 is reached, the calculations can be carried out as follows.

1. The ground distance, s_1 can be evaluated from Eq. 6.11 or 6.17 as outlined in Section 6.2.1 with velocity V_1.
2. The ground distance s_2 from V_1 to V_T is obtained from Eq. 6.11, with the augmented thrust-weight ratio included in

$$c = 2g\left(\frac{T_o}{W} + \frac{T_{aug}}{W} - \mu\right)$$

3. The time spent during the augmentation between V_1 and V_T can be evaluated from Eq. 6.12, with a and b determined in step 2 above, or by use of the average value of the time integral

$$t_2 = \frac{W}{g}\left(\frac{V_T - V_1}{T - \mu W - D + \mu L}\right)$$

where D and L are obtained at V_{aug} and T is the total augmented thrust.

4. The time t_2, obtained in step 3, is now compared with the firing time of the augmenting unit. If $t_2 < t_{aug}$ then a new and smaller

V_1 must be chosen and the steps 1 through 3 repeated. If $t_2 > t_{aug}$ then a larger V_1 is used for the next iteration step. The iteration can be repeated until t_2 and t_{aug} agree within satisfactory limits.

Effect of Wind The effect of wind is to reduce the ground run distance and time if the take-off is into a head wind. With take-off into downwind, both distance and time are appreciably increased. To establish the effect of wind it is necessary to calculate the distance traveled with respect to ground, as the distance traveled through air is not the same due to relative motion of air with respect to ground.

The ground speed V is then

$$V = V_a \pm V_w \tag{6.40}$$

where

V_a is airspeed
V_w is wind velocity [$-$ head wind, $+$ tail wind] (assumed to be constant)

The ground distance can be obtained by first establishing the acceleration and then integrating.

Now for the head wind, the acceleration a_g is

$$a_g = V\frac{dV}{ds} = (V_a - V_w)\frac{dV_a}{ds} = \frac{dV_a^2}{2ds} - V_w\frac{dV_a}{ds} \tag{6.41}$$

and

$$ds = \frac{dV_a^2}{2a_g} - \frac{V_w}{a_g}dV_a \tag{6.42}$$

Since $a_g = \dfrac{T - D - \mu(W - L)}{W/g}$

one obtains

$$s = \int_{\pm V_W}^{V_T}\frac{V\,dV}{a_g} - V_w\int_{\pm V_W}^{V_T}\frac{dV}{a_g} \tag{6.43}$$

and by use of Eq. 6.9

$$s_g = \frac{1}{2a} \ln \frac{aV_T^2 + bV_T + c}{aV_w^2 \pm bV_w + c} + \frac{b/a \pm 2V_w}{\sqrt{b^2 - 4ac}}$$
$$\times \left(\tanh^{-1} \frac{2aV_T + b}{\sqrt{b^2 - 4ac}} - \tanh^{-1} \frac{\pm 2aV_w + b}{\sqrt{b^2 - 4ac}} \right) \quad (6.44)$$

where the top sign is used for the head wind, and the bottom sign for downwind take-off.

NOTE

A convenient, but rather approximate, expression for the ground distance with the wind can be obtained from

$$d_s = (V_a \mp V_w) \frac{dV_a}{a_g} \quad (6.45)$$

which integrates into

$$s = \frac{(V \mp V_w)^2}{2a_g} \quad (6.46)$$

For transition and climb portions, the effect of headwind may be obtained approximately by multiplying the distance obtained in still air (Section 6.2.2) by the factor

$$\frac{(V_f \mp V_w)}{V_f} \quad (6.47)$$

where the signs stand for

− headwind
+ tailwind

and

$V_f = \dfrac{V_T + V_a}{2}$, for the transition phase

$V_f = V_a$, for the climb phase

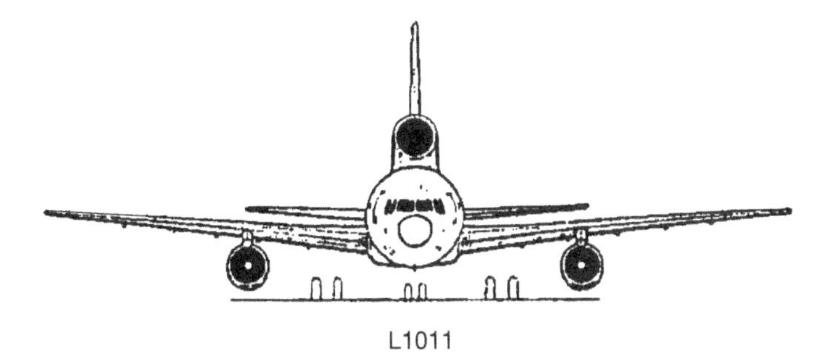

L1011

EXAMPLE 6.3

For Example 6.1, determine the take-off distance with 15 knot head wind. With head wind, the ground run is calculated from Eq. 6.44, the coefficients a and b remaining the same. Thus, $V_w = 25.35$ ft/sec, $V_w^2 = 643$ ft^2/sec^2

$$s = \frac{-10^5}{2 \times 2.02} \ln \frac{-2.02 \times 10^{-5} \times (164)^2 - 4.225 \times 10^{-3} \times 164 + 5.796}{-2.02 \times 10^{-5} \times 643 - 4.225 \times 10^{-3} \times 25 + 5.796}$$

$$+ \left(9500 + \frac{2 \times 25.35}{0.022}\right)\left[-0.5391 - \tanh^{-1} \frac{-2.02 \times 10^{-5} \times 2 \times 25.35 - 4.225 \times 10^{-3}}{0.022}\right]$$

$$= 5426 - 3492 = 1934 \text{ ft}$$

The effect of head wind on transition distances can be estimated by determining

$$V_f = \frac{V_T - V_a}{2} = 167.5 \frac{\text{ft}}{\text{sec}}$$

and calculating

$$s_t = \frac{V_f - V_w}{V_f} = \frac{167.5 - 23.35}{167.5} \times 320 = 272 \text{ ft}$$

Climb phase distance is estimated by letting $V_f = V_a = 170.9$ ft/sec and from the last part of the example:

$$s_c = \left(\frac{170.9 - 25.35}{170.9}\right) \times 458 = 390 \text{ ft}$$

by summing up, the total take-off distance is found to be, with the head wind,

$$s_{T.O._w} = s_g + s_r + s_t + s_c = 1934 + 492 + 272 + 390 = 3{,}088 \text{ ft}$$

6.3 LANDING

6.3.1 Landing Phases

Similarly to take-off operations, landing of a conventional aircraft should be divided into separate phases, as indicated in Figure 6.3:

1. The final approach, or gliding flight, at a relatively steady speed and rate of descent.
2. The flare, or flattening out transition phase. Here the pilot attempts (ideally) to reduce the rate of sink to zero and the forward

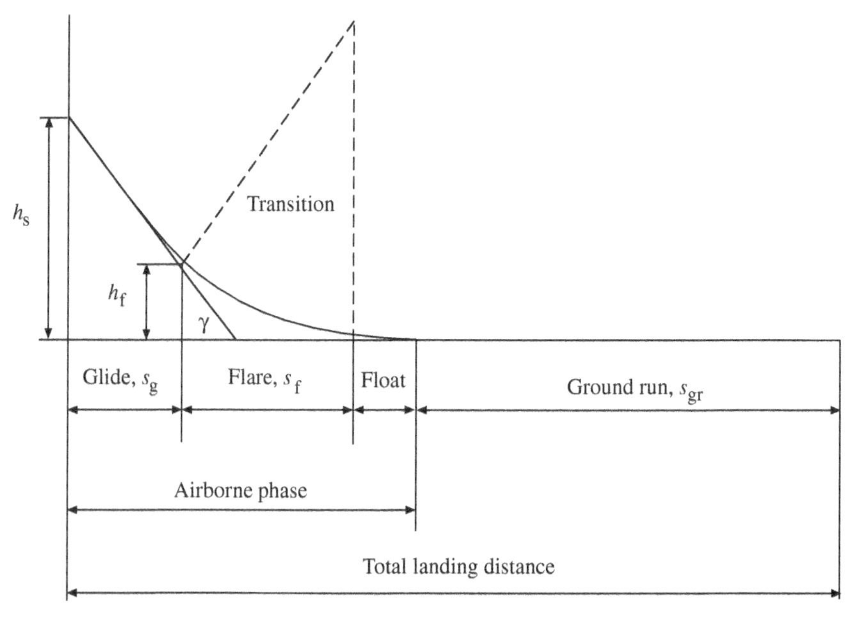

Figure 6.3 Total Landing Distance

speed to a minimum, which must be larger than V_s. To assure sufficiently high landing speed, Federal Air Regulations require that the velocity at screen height must be at least 1.3 V_s and 1.15 V_s at touchdown.

3. The floating phase, which is a result if the flare of zero rate of descent is achieved at the end of flare and the velocity must yet be reduced. The float is often eliminated for aircraft with nose-wheel type of landing gear, which, due to a relatively low ground angle of attack, are able to make a touchdown at speeds well above the stall speed. In general, the float occurs when the aircraft becomes subject to ground effect, which requires speed reduction for touchdown as the lift is now somewhat increased.
4. The ground run. In many cases, the ground run should be calculated in two sections. First, right after touchdown, the free rolling distance needs to be calculated while the pilot gets the nose wheel on the ground and starts applying brakes and reverse thrust. Second, the relatively longer deceleration distance must be determined that is required for the aircraft to come to rest or to reach a speed sufficiently low to be able to turn off the runway.

In practice, similar to take-off, piloting skill and varying techniques, different aerodynamic characteristics, and other factors such as differences of height and path angle all reduce the reliability of the calculated landing results. Inasmuch as take-off occurs under full power and at increasing speed, it is more repeatable. Landing, however, takes place under partial power and on the back side of the power curve. Thus, it is inherently less stable and pilots tend to find safety in speed. Ten pilots, in the same aircraft under same conditions, will produce ten different landings. Thus, the calculated final landing distance is multiplied by 1.67 to obtain a safe landing length. In case of wet runway, this distance is increased by another 15 percent.

Owing to the uncertainties just discussed, airborne phase of landing does not lend itself to a practical and reliable analysis. The ground phase, however, is quite significant for aircraft-airport compatibility (i.e., is there a sufficiently long runway to be able to land this particular aircraft?). Thus, more attention is devoted here to the landing run, and the airborne phase is included later only as a matter of completeness.

6.3.2 Landing Run

The landing ground run analysis is identical to the take-off run, and Eq. 6.9 can be used with $b = 0$, $C = 0$; see Eq. 6.6 and 6.8:

$$s_{gr} = \frac{-1}{2a} \ln \left(\frac{c + aV_T^2}{c + aV_2^2} \right) \tag{6.48}$$

Where V_T is now the touchdown velocity. V_2, in general, will be zero, and

$$c = g \left(\frac{T_o}{W} - \mu \right) \tag{6.49}$$

$$a = \frac{-S\rho g}{2W} (C_D - \mu C_L) \tag{6.50}$$

The essential differences are found in the coefficient c being possibly negative due to reverse thrust, braking parachute, or due to the fact that μ is now much larger than for take-off when brakes are used for a portion of the landing run. When reverse thrust is applied, then the thrust term T_o/W becomes negative in the constant c. Since it is unlikely that a pilot will apply brakes at the beginning of the ground run, for fear of getting into skid, the ground run may have to be divided into two or more portions to account for initial rolling friction, and then for the full brake section (usually at speeds below 80 knots). When the aircraft has been brought to rest, the landing run is determined by

$$s_{gr} = \frac{1}{2a} \ln \left(1 + \frac{a}{c} V_T^2 \right) \tag{6.51}$$

where C_D (especially C_{D0}) must be evaluated with all the aerodynamic braking devices and landing gear effects taken into account. C_{L_T} is obtained from touchdown speed V_T, which is between 1.15 and 1.2 V_s.

To account for the free rolling distance during which the pilot brings the nose wheel to the ground, one estimates similarly to take-off distance:

$$s_r = V_T \Delta t \tag{6.52}$$

where Δt is taken to be about 2 to 3 seconds.

The total calculated landing distance is now obtained from

$$s_{land} = s_{gl} + s_f + s_r + s_{gr} \tag{6.53}$$

which is given by Eqs. 6.57, 6.62, 6.51, and 6.52, or equivalent. As pointed out, the safe landing distance on a dry runway is obtained from

$$s = 1.67s_1 \tag{6.54}$$

NOTE

Some newer jets have a system called Autobrake, where the pilot can enter the desired deceleration during the ground run (after the nose wheel has been brought to the ground at a velocity V_1). Since the sensors establish a constant deceleration, call it D_1, the ground run can easily be determined from Eq. 6.4 as follows. The aircraft acceleration ($a = F/m$) is

$$a = \frac{T - D - \mu(W - L)}{W/g} \equiv D_1 \tag{6.55}$$

Thus, Eq. 6.4 can be integrated to give

$$s_r = \int_1^2 \frac{V\,dV}{D_1} = \frac{V_1^2}{2D_1} \tag{6.56}$$

Thus, if an aircraft makes contact with the ground at 160 ft/sec and the autobrake is set at 13 ft/sec^2, the expected ground distance is

$$s_{gr} = \frac{160^2}{2 \times 13} = 985 \text{ ft}$$

6.3.3 The Approach Distance*

The approach, or glide, distance can be estimated by assuming a straight glide path which gives

$$s_{gl} = \frac{h_s - h_f}{\gamma} \tag{6.57}$$

where γ is a small angle and h_s is the screen height. Then

$$\tan \gamma = \frac{D - T}{L} = \frac{1}{L/D} - \frac{T}{L} \tag{6.58}$$

where L/D is usually evaluated at a mean speed between 1.3 V_s, and 1.15 V_s, or simply assumed to be 8. For a first approximation, the thrust term is often neglected; however, a better procedure is to use the idle thrust value. The glide velocity is assumed to be constant at 1.3 V_s.

6.3.4 The Flare Distance*

Estimation of the flare distance is the least reliable of the landing distance calculation procedures, as it requires a number of assumptions concerning the shape of the path, velocities, start and end points, and so on.

Thus, there are many methods and variations that define the arc used to calculate the flare distance. They all give similar results based on a variety of reasonable physical assumptions that are judiciously tempered by experimental observations. In what follows, two approaches are given. The first one, Method A, is included as a typical example to indicate the basic methodology. A thoughtful observer has no difficulty raising more questions than are answered here. Method B is based on the energy conservation and seems to yield results comparable to other more elaborate methods. Its main advantages lie in simplicity and the individual number of assumptions involved.

Method A The following steps and assumptions are usually involved:

1. The flare path is a circular arc of such a radius that the normal acceleration is 0.1 g.
2. The flare is carried out at a constant speed:

$$V_{\mathrm{f}} = \frac{1.3V_{\mathrm{s}} + 1.15V_{\mathrm{s}}}{2} = 1.23V_{\mathrm{s}}$$

Thus, the radius of the circle is (see Figure 6.3)

$$r_{\mathrm{f}} = \frac{(1.23)^2 V_{\mathrm{s}}^2}{0.1g} = 15.1\,\frac{V_2^2}{g}$$

Furthermore, it is assumed that the flight angle is sufficiently small that the horizontal distance is equal to the arc length. Then, if the flare height h_{f} is known, or assumed:

$$s_{\mathrm{f}} = 15.1\,\frac{V_{\mathrm{s}}^2 \gamma}{g}$$

where γ can be evaluated from the following (see Figure 6.4):

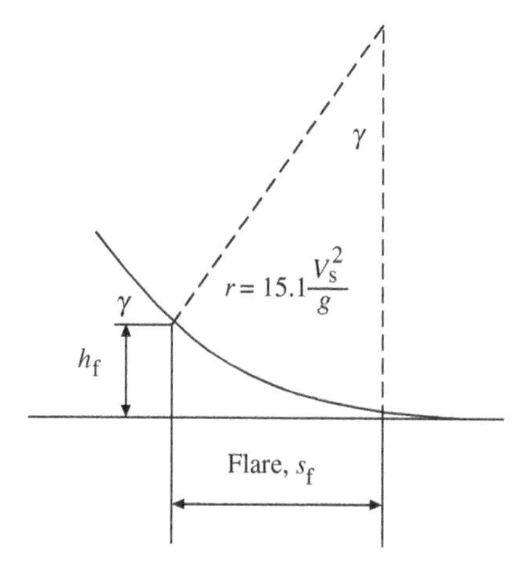

Figure 6.4 Transition Geometry

$$\cos\gamma = \frac{15.1V_s^2 g - h_f}{15.1V_s^2 g}$$

Otherwise, it may be assumed that γ is determined by L/D ($\equiv 8$). If the flare starts at a height above the screen height h_s, then

$$s_f = \frac{15.1V_s^2}{g}\gamma + \frac{h_s - h_f}{\tan\gamma} \tag{6.59}$$

Method B Considering now the uncertainties involved in evaluating the glide and flare distances, one might as well calculate the entire airborne distance by considering the energy equation derived in Chapter 2. For simplicity, the energy equation can be applied between the screen height and the touchdown point. Thus, one obtains with Eq. 2.15

$$We_s - We_T = D_{S_a} - T_{S_a} \tag{6.60}$$

Upon substituting kinetic and potential energies at screen height and at landing ($h = 0$), one obtains:

$$W\left(h_s + \frac{V_{sc}^2}{2g} - \frac{V_T^2}{2g}\right) = s_a(D - T)$$

or

$$W\left(h_s + \frac{(1.3)^2V_s^2}{2g} - \frac{(1.15)^2V_s^2}{2g}\right) = s_a(D - T) \qquad (6.61)$$

where

s_a is the airborne distance
e_s is the specific energy at screen height
e_T is the specific energy at touchdown, $(h = 0)$
V_{sc} is the velocity at screen height
V_T is the touchdown velocity
V_s is the stall speed

Then the airborne distance is obtained as

$$s_a = \frac{W}{D - T}\left(h_s + \frac{0.3675V_s^2}{2g}\right) \qquad (6.62)$$

where D must be evaluated at an average speed V_f corresponding to a lift that yields an $L/D \equiv 8$. Often, thrust is assumed to be zero. A much better approach is to use engine idle thrust value, which amounts to about 6 to 9 percent of rated thrust.

6.4 ACCELERATING FLIGHT*

In level flight, an aircraft accelerates or decelerates every time its thrust (or speed) is changed. Since most of the flight takes place at relatively constant velocity, the problems involving accelerations are of interest near beginning or end of the flight or arise due to special maneuvers—for example, in combat. In addition, many aircraft are equipped with air brakes to provide speed control in dive or combat, or to assist in flight path control on landing approach. The essential problems are finding the time and distance to accomplish the speed change and to evaluate the effectiveness of the air brakes.

Consider an aircraft in steady, level flight with velocity V_1 at an altitude h (Figure 6.5).

At this flight condition, the drag is exactly balanced by the thrust:

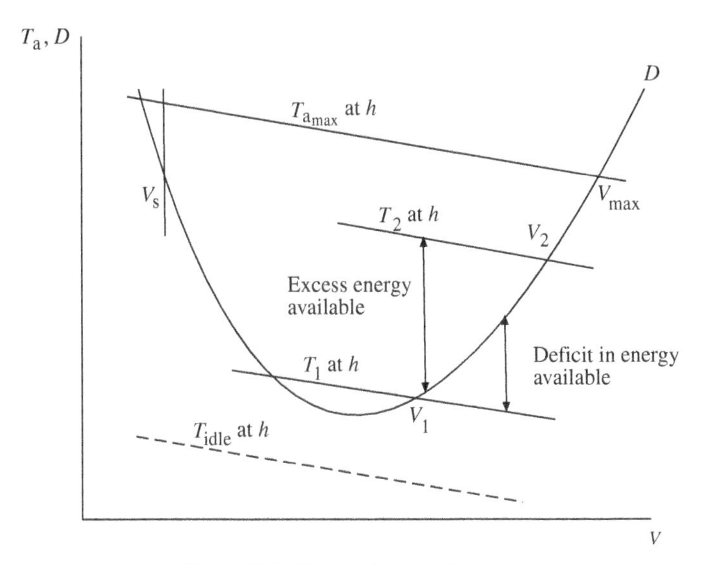

Figure 6.5 Energy for Acceleration

$$D_1 = T_1 < T_{max_{avail}}$$

An increase in the thrust, at the given altitude h, accelerates the aircraft from V_1 up to any desired velocity $V_2 \leq V_{max}$ depending on the level of thrust T_2. Excess energy $T_a - D$ will be available to accelerate the aircraft until the maximum velocity V_{max} is reached where the available thrust exactly balances the drag force. Acceleration beyond V_{max} may be possible if it is accompanied by a change in altitude. Conversely, if the aircraft is in steady level flight at V_2 and the thrust is reduced to T_1, the excess of drag energy (deficit in energy available) will serve to decelerate the aircraft to V_1. Suppose now that the thrust is further cut to its idle value T_i, then the aircraft will continue decelerating to reach stall velocity V_s if constant altitude can be maintained according to $L = W$, or $C_L V^2 =$ constant. In general, it cannot, as shown in Figure 6.5.

The general solution of accelerating flight requires integration of the dynamic Eqs. 2.9 through 2.13 and requires a complete description of the flight path, as discussed in Chapter 1. However, if the problem is approached from the point performance point of view, or the equations of motion are integrated by the use of simplifying assumptions, some practical solutions can be obtained.

Eqs. 2.9 through 2.12 reduce, for level flight or for flight with small path angle change, to

$$\frac{dx}{dt} = V \tag{6.63}$$

$$\frac{1}{g}\frac{dV}{dt} = \frac{T - D}{W} \tag{6.64}$$

$$L = W \tag{6.65}$$

It is expedient now to change the independent variable to velocity V, as carried out in Section 6.2.1, and these equations become

$$dx = \frac{W}{g}\frac{V\,dV}{T - D} = \frac{V\,dV}{a} \tag{6.66}$$

$$dt = \frac{W}{g}\frac{dV}{T - D} = \frac{dV}{a} \tag{6.67}$$

Eq. 6.65 reduces to stating that, for level flight, $C_L V^2$ = constant, which implies that the angle of attack must be changed whenever the velocity is changed.

In point performance approach, the following numerical integration procedure can be used (for either acceleration or deceleration):

1. Initial and final velocities are assumed.
2. At selected velocity intervals calculate
   $C_L = \frac{\text{const}}{V^2}$
   $C_D = f(C_L)$, from a given drag polar
   T, from engine data at selected V or h (usually assumed to be constant)
3. Evaluate, at selected V

$$\frac{1}{a} = \frac{w}{(T - D)g} \quad \text{and} \quad \frac{V}{a}$$

4. Plot $1/a$ and V/a vs. V
5. The areas under the curves in Figure 6.6 give the time or distance in accelerated flight.

At small flight path angles, the above process will provide useful approximations to climbing or gliding flight, provided a consistent altitude/velocity program is chosen.

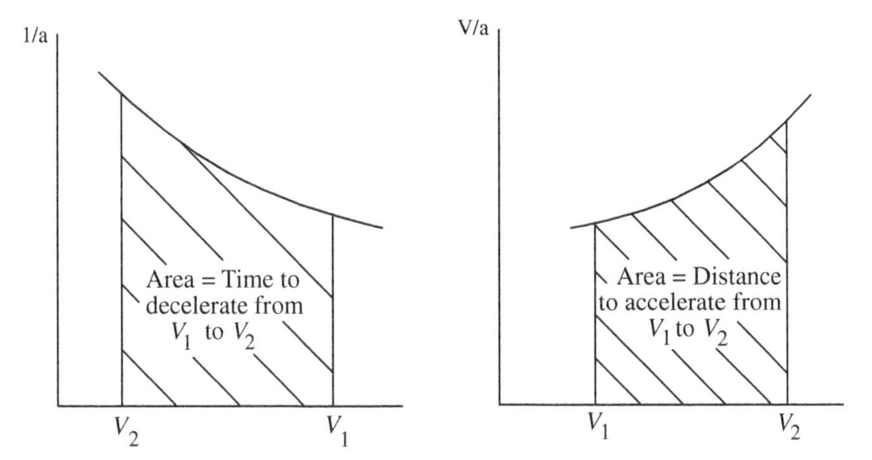

Figure 6.6 Time and Distance Calculations

The effect of speed brakes is evaluated by including the additional drag due to speed brakes in calculating the drag coefficient C_D.

EXAMPLE 6.4

An aircraft weighs 105,000 N, has a wing area of 75 m^2 and is traveling at 100 m/s at 6100 m altitude. Estimate the time to accelerate this aircraft to a speed of 250 m/sec if its thrust is increased to a constant value of 40,000 N.

$$C_D = 0.018 + 0.055C_L^2$$

The calculations are carried out for initial and final velocities of 100 and 250 m/s, respectively, at 50 m/s intervals. The lift coefficient is evaluated from

$$C_L = \frac{2W}{\rho S V^2} = \frac{2 \times 105{,}000}{1.2256 \times 0.5383 \times 75 \times V^2} = \frac{4{,}244}{V^2}$$

The drag is calculated by $D = W(C_L/C_D)$, whence

$$a = \frac{T - D}{W/g} = \frac{40{,}000 - D}{105{,}000/9.807} = \frac{40{,}000 - D}{10{,}707}$$

Calculations at 50 m/s intervals produce Table 6.2.

TABLE 6.2 Example 6.4 Data

V	100	150	200	250
C_L	0.424	0.189	0.106	0.068
C_D	0.0279	0.02	0.0186	0.0183
D	6909	11,091	18,442	28,187
a	3.09	2.7	2.01	1.1
$1/a$	0.324	0.37	0.497	0.906

A straightforward numerical integration gives the result that it takes about 75 seconds to accelerate from 100 to 250 m/sec. (See also Problem 6.9).

Eqs 6.66 and 6.67 can be integrated directly to give time and distance if it is assumed that

1. The thrust T is independent of velocity (is essentially assumed to be constant).
2. The simple drag polar

$$C_D = C_{D_0} + kC_L^2$$

is applicable.

Integration is facilitated by introducing the following nondimensional variables:

$$\bar{x} = \frac{xg}{V_{D\min}^2 E_m}$$

$$\bar{t} = \frac{tg}{V_{D\min}^2 E_m}$$

$$\bar{V} = \frac{V}{V_{D\min}}$$

$$\bar{T} = \frac{T}{W} E_m$$

Eqs. 6.66 and 6.67 become

$$d\bar{x} = -\frac{2\bar{V}^3 d\bar{V}}{\bar{V}^4 - 2\bar{T}\bar{V}^2 + 1} \tag{6.68}$$

$$d\bar{t} = -\frac{2\bar{V}^2 d\bar{V}}{\bar{V}^4 - 2\bar{T}\bar{V}^2 + 1} \tag{6.69}$$

Integration now leads to

$$x = -\frac{V_{D\min}^2 E_m}{g}\left[\frac{1}{2}\ln(\bar{V}^4 - \bar{T}\bar{V}^2 + 1) + \bar{T}Q\right] + \text{constant} \tag{6.70}$$

where

$$Q = \frac{1}{2\sqrt{\bar{T}^2 - 1}} \ln \frac{\bar{V}^2 - \bar{T} - \sqrt{\bar{T}^2 - 1}}{\bar{V}^2 - \bar{T} + \sqrt{\bar{T}^2 - 1}}, \text{ if } \bar{T}^2 - 1 > 0$$

$$Q = \frac{1}{2\sqrt{1 - \bar{T}^2}} \tan^{-1} \frac{\bar{V}^2 - \bar{T}}{\sqrt{1 - \bar{T}^2}}, \text{ if } \bar{T}^2 - 1 < 0$$

$$Q = -\frac{1}{\bar{V}^2 - \bar{T}}, \text{ if } \bar{T}^2 - 1 = 0$$

and the expression for time becomes

$$t = \frac{V_{D\min} E_m}{g(\bar{V}_2^2 - \bar{V}_1^2)}\left[\bar{V}_2 \ln \frac{\bar{V} + \bar{V}_2}{\bar{V} - \bar{V}_2} - \bar{V}_1 \ln \frac{\bar{V}_1 + \bar{V}}{\bar{V}_1 - \bar{V}}\right] + \text{constant, if } \bar{T}^2 - 1 > 0 \tag{6.71}$$

where

$$\bar{V}_1 = \bar{V}_{max} = \sqrt{\bar{T} + \sqrt{\bar{T}^2 - 1}}$$

$$\bar{V}_2 = \bar{V}_{min} = \sqrt{\bar{T} - \sqrt{\bar{T}^2 - 1}}$$

and

$$t = -\frac{V_{D_{\min}} E_{\mathrm{m}}}{g}\left[\frac{\sin\beta}{2\sqrt{1-\overline{T}^2}}\ln\frac{\overline{V}^2+1-2\overline{V}\cos\beta}{\overline{V}^2+1+2\overline{V}\cos\beta}\right.$$

$$\left.+\frac{\cos\beta}{\sqrt{1-\bar{t}^2}}\left[\tan^{-1}\frac{\overline{V}-\cos\beta}{\sin\beta}+\tan^{-1}\frac{\overline{V}+\cos\beta}{\sin\beta}\right]\right]$$

$$+\text{ constant, if } \overline{T}^2-1<0 \tag{6.72}$$

where

$$\beta = \frac{1}{2}\tan^{-1}\frac{\sqrt{1-\overline{T}^2}}{\overline{T}}.$$

It should be pointed out that the $\tan^{-1}$ terms should not be added, as is often a customary procedure. There is a hidden singularity at $\overline{V} = 1$.

EXAMPLE 6.5

Calculate the distance to decelerate for an aircraft, which is decelerating at sea level from initial velocity $V_{\mathrm{i}} = 600$ ft/sec to final velocity $V_{\mathrm{f}} = 300$ ft/sec at idle thrust of 250 lb. The aircraft weighs 18,000 lb and has a clean drag polar for a wing area of 300 ft^2.

$$C_D = 0.014 + 0.06C_L^2$$

Upon calculating

$$E_{\mathrm{m}} = \frac{1}{2\sqrt{0.014 \times 0.06}} = 17.25$$

$$\overline{T} = \frac{T}{W}E_{\mathrm{m}} = \frac{250 \times 17.25}{18{,}000} = 0.24$$

and since

$$\overline{T}^2 - 1 < 0$$

one finds the proper form of Eq. 6.70 to be

$$x = \frac{V_{D_{\min}}^2 E_m}{g}\left[\frac{1}{2}\ln\frac{\overline{V}_i^4 - 2\overline{T}\overline{V}_i^2 + 1}{\overline{V}_f^4 - 2\overline{T}\overline{V}_f^2 + 1} + \frac{\overline{T}}{\sqrt{1 - \overline{T}^2}}\right.$$

$$\left.\left[\tan^{-1}\frac{\overline{V}_i^2 - \overline{T}}{\sqrt{1 - \overline{T}^2}} - \tan^{-1}\frac{\overline{V}_f^2 - \overline{T}}{\sqrt{1 - \overline{T}^2}}\right]\right]$$

Evaluating now

$$V_{D_{\min}} = \sqrt{\frac{2W}{\rho S}}\left(\frac{k}{C_{D_0}}\right)^{1/4} = \sqrt{\frac{2 \times 18{,}000}{0.002377 \times 300}}\left(\frac{0.06}{0.014}\right)^{1/4}$$

$$= 323\ \frac{\text{ft}}{\text{sec}}$$

$$\overline{V}_i = \frac{V_i}{V_{D_{\min}}} = \frac{600}{323} = 1.86$$

$$\overline{V}_f = \frac{V_f}{V_{D_{\min}}} = \frac{300}{323} = 0.929$$

The solution for the deceleration distance is found to be

$$x = \frac{(323)^2 \times 17.25}{32.2}\left[\frac{1}{2}\ln\frac{11.31}{1.33} + \frac{0.24}{0.971}\right.$$

$$\left.(\tan^{-1} 3.316 - \tan^{-1} 0.641)\right] = 69{,}589 \text{ ft}$$

PROBLEMS

6.1 For Example 6.1, evaluate ground run from Eq. 6.22 with $0.8V_T$. Ans: 2,285 ft.

6.2 For Example 6.1, evaluate ground run when the thrust reversal is used from $0.8V_L$ to 60 mph. Assume that the thrust reverses produce 50 percent of sea level rated thrust, and that lift and drag coefficients during the landing run are $C_L = 0.2$ and $C_D = 0.1$. Assume that no brakes are applied and that the idle thrust

is 5 percent of the rated sea level value.
Ans: 8,570 ft.

6.3 A twin jet aircraft is taking off from a concrete runway (μ = 0.04). The take-off polar is

$$C_D = 0.075 + 0.04C_L^2$$

with

$$S = 54.8 \text{ m}^2$$
$$\mathcal{R} = 6$$
$$W = 266{,}880 \text{ N}$$
$$C_{L_{\max}} = 2.7$$
$$\text{T/engine} = 80{,}064 \text{ N} = \text{const}$$
$$T_{\text{idle}} = 0.07T$$

Calculate:

a. The ground run

b. The ground run with one engine failing at 100 mph.

c. The head wind required to reduce the ground run in *b* to that obtained in *a*—that is, what head wind is required to compensate for the loss of one engine?

d. Time for *a*.

e. Time for *b*.

Ans: a. 400 m; b. 672 m; c. 24.4 m/s; d. 11.8 s; e. 17 s.

6.4 For Example 6.1, calculate the ground run with a 15 knot tail wind.
Ans: 2,994 ft

6.5 Evaluate the effect of C_{L_T}, W/S, and T/W on ground run distance for two aircraft, at W/S = 50 and 100. Assuming that the take-off polar for each is $C_D = 0.03 + 0.05C_L^2$, and that $\mu = 0.02$, plot the ground run distance vs. C_{L_T} with T/W and W/S as parameters. What conclusions can be drawn regarding increase of C_{L_T}?

6.6 A transport aircraft weighs 290,000 lbs and is equipped with four JT3D jet engines. It has the following characteristics:

$$C_{L_{max}} = 2.2$$
$$AR = 7.346$$
$$S = 2{,}892 \text{ ft}^2$$
$$C_{D_0} = 0.021$$

Calculate

a. Ground run, standard day, in Los Angeles

b. Ground run, 100°F, in Los Angeles

c. Ground run, standard day, Mexico City (7,400 ft elevation)

6.7 For Example 6.5, calculate the time to decelerate from 600 ft/sec to 300 ft/sec.
Ans: 2.7 min.

6.8 For Example 6.5, calculate the distance and time to decelerate from 600 ft/sec to 300 ft/sec if speed brakes are activated that have a drag coefficient of 1.0, based on the brakes area of 20 ft^2.
Ans: 2.5 mi, 32 s.

6.9 Verify the answer obtained in Example 6.4 by using the appropriate integrated result for time.

6.10 A twin-engine cargo plane is being loaded for takeoff. The aircraft has the following characteristics:

$$C_{D_{clean}} = .019 + .048C_L^2,\ S = 880 \text{ ft}^2$$
$$C_{LTO} = 2.1,\ C_{DTO} = .03 + .06C_L^2$$
$$W_{empty} = 59{,}000 \text{ lb (incl fuel and crew).}$$

Sea-level rated thrust is 10,000 lb/engine. Ignore μ.

Airport runway is rather short, translating into a ground run of about 2,000 ft, before rotation. It is a hot day with the air density being only 92 percent of that of the standard day, and the thrust available has thus dropped to 92 percent of the rated value.

The charts show that the maximum weight allowable is about 10,000 lb cargo. The crew chief insists that 13,000 lb is fine. Who is right?
Ans: The charts are right.

6.11 A large transport aircraft has the following characteristics:

$W_{TO} = 605{,}000$ lb
$S = 3{,}650$ ft^2
3 engines, 61,500 lb each, T_{max}
$C_D = .016 + .042C_L^2$, $C_{L_{max}} = 2.4$

Determine:

a. Lift-off velocity
b. The take-off roll
c. The time to lift off

6.12 An A7E has just landed. It takes 3 seconds to bring the nose gear down and to stabilize for the ground run. It has the following characteristics:

$T_a = 11{,}000$ lb
$C_D = .015 + .1055C_L^2$
$C_{Lmclean} = 1.6$
$C_{Lmfl,slats} = 2.5$
$W = 25{,}000$ lb, $S = 375$ ft^2
$T_{idle} = .09T_a$

Calculate the total distance covered on the ground if:

a. No brakes are applied at all.
b. Brakes are applied after the end of first three seconds (use $\mu = .04$).
Ans: about 190,000 ft; 2,100 ft.

6.13 The following is given for a twin-engine F18 aircraft:

$C_D = .0245 + .13C_L^2$, clean; add $\Delta C_{D_0} = .075$ for landing gear
$S = 400$ ft^2, $W = 40{,}000$ lb
$C_{L_{max}} = 2.8$; $C_{L_{TO}} = .2$
T = 9,300 lb/engine

Calculate the take-off ground run ($\mu = .03$) for:

a. The data given above
b. Take-off with full afterburners, 13,000 lb/engine
Ans: Using Eq. 6.13, 1,642 ft; 1,130 ft

7

Maneuvering Flight

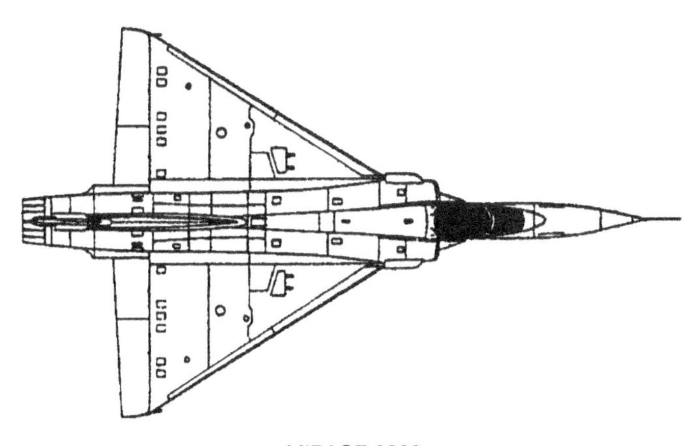

MIRAGE 2000

7.1 INTRODUCTION

In a steady, unaccelerated flight all the forces acting on an aircraft are in equilibrium. For about 90 percent of flight, the simple equilibrium condition $C_L V^2$ = constant holds, which governs simple and minor adjustments to the flight path. In point performance approach, this may be assumed even for modest rectilinear acceleration. This chapter considers flight where the path direction changes sufficiently rapidly so that the above condition may not be valid.

In order to accelerate or curve the flight path in any plane—that is, to pull up or to turn—the force equilibrium acting on the aircraft must become unbalanced. This can be accomplished essentially in three ways: change lift, change velocity, or bank the aircraft. The lift coefficient will be changed so rapidly that the velocity cannot keep up with the change and the equilibrium condition ($C_L V^2$ = const) will not be satisfied. The unbalanced force then acts to create an acceleration normal to the flight path. If a velocity change is the cause, it must be produced sufficiently rapidly so that C_L cannot compensate. The result is again a normal acceleration. In vertical plane (pull-ups, push-overs) rapid lift and velocity changes are the principal mechanisms for causing accelerations.

Accelerations out of the vertical plane (not necessarily normal to it, but normal to the flight path) can be caused by rolling the aircraft into a turn. The lift vector is then rotated out of the vertical plane by rolling the aircraft (banking the wings). This action produces a lift component to act as an accelerating force for a turn and leaves a portion of lift force to balance the weight vector.

If the vertical component of the lift force, usually by getting some assistance from a vertical component of thrust, does not exactly balance the weight force, then climb or descent will accompany the turning motion. The analysis of resulting three-dimensional turning performance is rather complicated, and the governing set of differential equations admits only a few solutions, which provide scant insight into the phenomena at hand.

Thus, in keeping with the general point performance approach taken above, the motion will be restricted to either vertical or horizontal plane and analytic solutions will be considered at distinct points of the flight path. The resulting equations should provide adequate tools for analyzing the essential turning performance of an aircraft.

7.2 TURNS IN VERTICAL PLANE: PULL-UPS OR PUSH-OVERS

To determine what turn rate may be obtained, or more important, what accelerations could be produced in a pull-up maneuver, consider Figure 7.1, where the lift generated now can be expressed as

$$L = nW \tag{7.1}$$

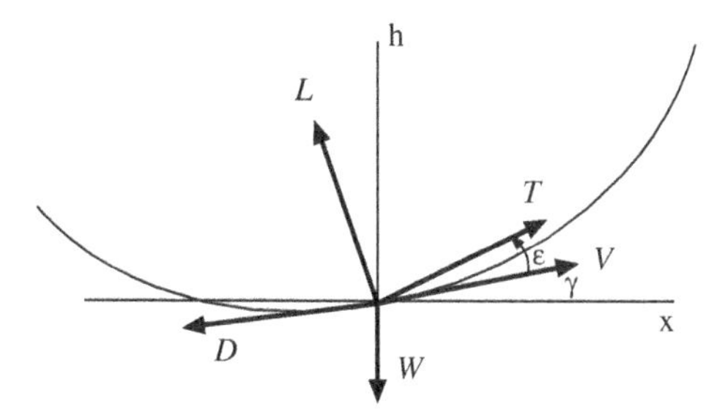

Figure 7.1 Pull-up Maneuver

The load factor n, usually expressed as g's of acceleration, is defined by Eq. 7.1 and represents the excess of lift generated over the weight to produce the required maneuver.

Since the flight path is in the vertical plane, Eqs. 2.5 and 2.6 apply directly. As the maneuver is performed very rapidly, the flight speed remains instantaneously unchanged and those equations uncouple. Thus, only Eq. 2.6 needs to be considered.

$$L + T \sin \epsilon - mg \cos \gamma = mV\dot{\gamma}$$

which becomes with $L = nW$ and $mg = W$

$$\frac{g(n - \cos \gamma)}{V} + \frac{Tg \sin \epsilon}{WV} = \dot{\gamma} \tag{7.2}$$

and is usually simplified to give

$$\dot{\gamma} = \frac{g(n - 1)}{V} \tag{7.3}$$

Eq. 7.3 gives the rotation rate for a modest thrust-to-weight ratio aircraft at the beginning of pullout from level flight. Since T/W ratio for high-performance aircraft approaches, or even exceeds, unity, Eq. 7.2 shows that thrust contribution may be significant and should not always be ignored.

Centrifugal acceleration may now be determined easily by recalling that $a = \dot{\gamma}V$, whence

$$a = g(n - \cos \gamma) + \frac{Tg \sin \epsilon}{W} \approx g(n - 1) \tag{7.4}$$

Eq. 7.4 indicates the acceleration on the pilot and aircraft in an n "g" pull-out. It should be recognized that acceleration has the same effect on the pilot and the structure: to change the direction of the flight path. The pilot, however, senses the load factor only in excess of unity. At $n = 1$ (level flight) the pilot feels no load in excess to that felt when sitting at home on the couch. At $n = 1$, the wings already feel the load, weight W, of the aircraft, and at $n = 2$ must be able to generate lift capable of supporting $2W$. Thus, as the load factor increases, so does the load acting on the aircraft. But same load factors do not necessarily produce same loads as the weight of the aircraft changes during flight and its missions. A load factor of 5 may be just within safe operating load limits of an aircraft carrying only a part load of fuel. When fully fueled and with external stores attached, the total weight could increase by 50 percent. Now a load factor of 5 may require wing loads that could have disastrous structural consequences.

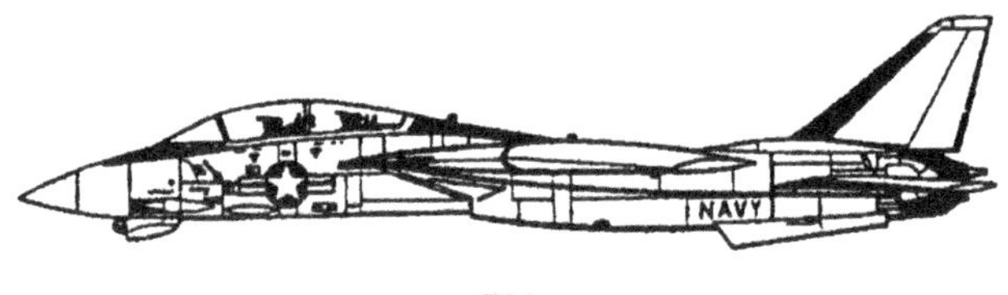

F14

7.3 V–*n* DIAGRAM

The operation of an aircraft is subject to rather specific operating strength limitations, usually presented on what is known as the V–n or V–g diagram. The basic static strength requirement is stated as the positive limit load factor n_l. This means that an aircraft, in its design or mission configuration, is expected to meet a peak load corresponding to limit load factor n_l and to withstand this load without permanent deformation of the structure. To allow for situations where mission or emergency situations may lead to exceeding the limit load factor, the ultimate load factor 1.5 n_l is provided. At this load, the primary load-carrying structure may exhibit some permanent deformations but no actual failures should take place. Some typical values for the limit load factors are listed below. These values are only typical, as they may change with specific mission requirements.

Type of Aircraft	Limit Load Factor
Fighter, Trainer, Attack	7–9
Transport, Patrol	3
Passenger	3–4

The above load factors provide only a bound for loads that must be observed carefully during aircraft operation and may easily be exceeded if due regard is not given to the operating limits that must be established from additional considerations.

The loads on an aircraft are produced in two ways. The *intentional* loads are due to maneuvers and the *unintentional* ones are due to air disturbances—gusts. In attack and fighter aircraft design and performance analysis, the maneuver loads are more important. In passenger aircraft operations, where operating load factors seldom exceed 1.5, the gust loads are more significant. Although this chapter is essentially concerned with maneuvers, the gust-originated load factors will also be considered due to their impact on high-speed flight at lower altitudes.

The load factor can be related to velocity by dividing the instantaneous values of lift and weight. Of importance is the maximum load factor, which is obtained from maximum lift at any airspeed:

$$L_{\mathrm{max}} = \frac{1}{2} C_{L_{\mathrm{max}}} \rho V^2 S \tag{7.5}$$

At stall speed, maximum lift is equal to the weight:

$$W = \frac{1}{2} C_{L_{\mathrm{max}}} \rho V_{\mathrm{s}}^2 S \tag{7.6}$$

From definition of the load factor, then

$$n_{\mathrm{max}} = \frac{L_{\mathrm{max}}}{W} = \left(\frac{V}{V_{\mathrm{s}}}\right)^2 \tag{7.7}$$

Eq. 7.7 shows that the load factor varies as the square of the aircraft velocity (ratio of the stall speed). Thus, if the aircraft is flying at the stall speed, load factor is unity. At twice the stall speed, a load factor of 4 can be obtained if the angle of attack is rapidly increased to produce maximum lift. At four times the stall speed, a load factor of 16 will result, which exceeds the structural capabilities of all aircraft, with the possible exception of special-purpose research planes. It follows, then, that high-speed aircraft, which fly at several times the stall speed, are capable of g loads that will exceed all structural limits. To prevent exceeding these limits, safe flight boundaries are presented on a V–n diagram, Figure 7.2.

The horizontal boundaries represent the limit load factors as defined for this particular aircraft. Also shown are the ultimate load factors.

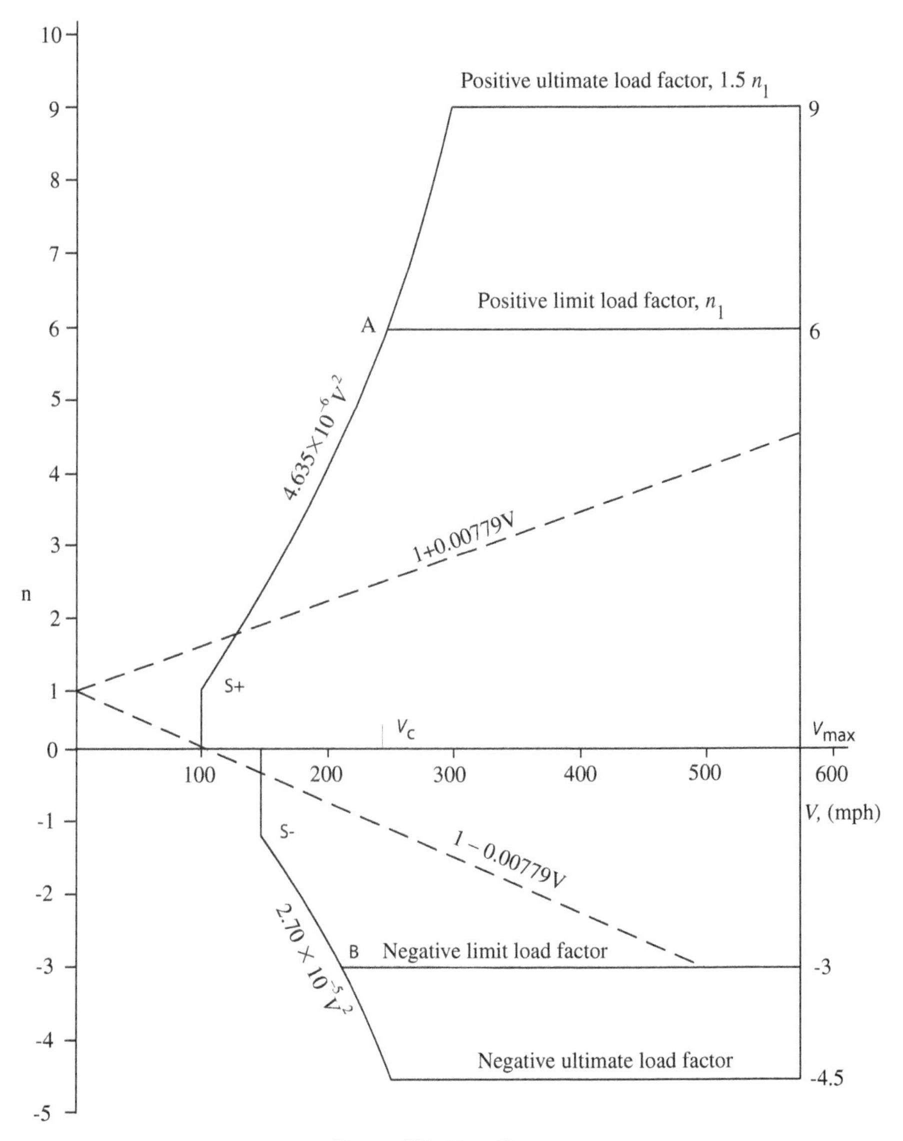

Figure 7.2 V–n Diagram

The curved boundaries S–A and S–B are obtained by means of Eq. 7.7; $V_S = 100$ mph where $n = 1$. S–A and S–B can also be obtained from Eq. 7.5, as follows:

Consider an aircraft with the following sea level characteristics.

$S = 650 \text{ ft}^2$
$W = 20{,}000$ lb
$+C_{L_{\max}} = 1.2$
$-C_{L_{\max}} = 0.7$
$a = 0.08$ per deg (both positive and negative lift)
$V_{\max} = 570$ mph, at sea level
$+n_{1_{\max}} = 6$
$-n_{1_{\max}} = 3$
$V_s = 100$ mph $(+C_{L_{\max}})$
$V_s = 131$ mph $(-C_{L_{\max}})$

Then the load factor can be expressed as

$$\begin{aligned} n &= \frac{1}{2}\frac{S}{W} C_{L_{\max}} \rho V^2 \\ &= \frac{1}{2}\left(\frac{650}{20{,}000}\right) \times 0.002377 \times 1.2 \times V^2 \\ &= 4.635 \times 10^{-5}\, V^2 \text{ for positive lift} \\ &= -2.704 \times 10^{-5}\, V^2 \text{ for negative lift} \end{aligned} \tag{7.8}$$

When evaluating n from Eq. 7.8, care should be exercised in evaluating $C_{L_{\max}}$. It may not be sufficient to use the static $C_{L_{\max}}$ value as presented in standard tables or charts. First, under dynamic loading conditions when a wing is rotated rapidly to a high angle of attack higher maximum lift conditions than for the static loading may be obtained for brief periods. Thus, higher load factors can be developed than predicted by static stall speed. Second, the lift-generating capability of the tail of some aircraft may be of sufficiently high magnitude (10 to 20 percent) that it should be included in evaluating $C_{L_{\max}}$ for calculating the load factor n.

Eq. 7.8 shows that as the altitude and weight change, a new diagram must be constructed. In addition, $C_{L_{\max}}$ and S may also change depending on configuration, external stores, and so on. Eq. 7.7 remains valid in any case, since the altitude and weight changes can be represented as a change in V_s, which then only requires shifting the velocity scale.

The upper, positive, n scale implies that a pull-up maneuver is performed. Some aspects of turning maneuvers will later be superimposed on that portion of the diagram. The lower, negative scale is used for negative accelerations in a push-over maneuver. The n values are smaller because the values of negative $C_{L_{\max}}$ are, in general, smaller (in this case -0.7) than the normal positive $C_{L_{\max}}$. Since a pilot's ability to withstand negative max g's is more limited than in the case of positive values, aircraft normally are designed for smaller values of negative load factors, as shown in Figure 7.2. The vertical limit defining the maximum speed is calculated from the terminal dive speed or is determined by compressibility considerations or from buffeting boundary.

The superimposed dashed lines are determined by calculating the aircraft response from anticipated gust loads. Gusts are horizontal and vertical velocity disturbances found in the atmosphere. In general, gusts are rarely found in stratosphere but occur frequently in the lower atmosphere in the vicinity of thunderstorms, mountains, shorelines, and in areas where thermal gradients exist. Crossing the wake of another aircraft also has the effect of a gust.

A horizontal gust causes a change in dynamic pressure over the wing, which does result in change of lift, but the load factors produced are relatively small and they can be ignored. A horizontal gust does not change the angle of attack. A vertical gust causes a change in angle of attack, as shown in Figure 7.3, and a change in load factor.

A gust consists of a wave-shaped velocity variation with low velocities at extremes and a maximum near the center. For analytical purposes, it is customary to assume that a gust behaves as a sharp-edged impulse and an equivalent velocity value of $KU = 30$ ft/sec is assumed to represent the strongest gust likely to be encountered in a normal operation. K is the airplane response factor (0.4–0.7) and U is the maximum gust velocity of 50 to 75 ft/sec. Thus, a *normal operation* combination of $KU = 30$ is assumed.

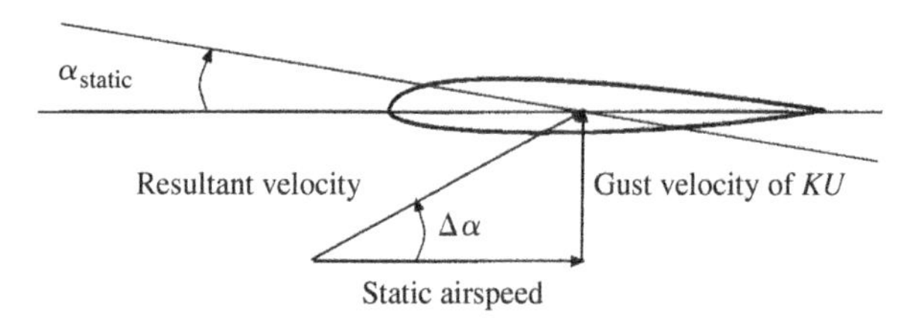

Figure 7.3 Effect of Gust

The effect on load factor can now be calculated by evaluating the changes in angle of attack and in the lift coefficient. For small $\delta\alpha$

$$\delta\alpha = \frac{KU}{V} \tag{7.9}$$

From definition of the lift curve slope

$$a = \frac{C_L}{\alpha} = \frac{\delta C_L}{\delta\alpha} \tag{7.10}$$

$$\delta C_L = a\delta\alpha = a\,\frac{KU}{V} \tag{7.11}$$

The change in lift is obtained as

$$\delta L = \frac{1}{2}\frac{KU}{V}\,\rho a V^2 S \tag{7.12}$$

and the change in load factor becomes

$$\delta n = \frac{\delta L}{W} = \frac{\rho}{2}\frac{KUV}{W/S}\,a \tag{7.13}$$

Since the aircraft is flying, before encountering the gust, at $n = 1$, the load factor can be written as

$$n = 1 + \frac{\rho a KUV}{2W/S} \tag{7.14}$$

For negative gusts the second term becomes negative as $KU < 0$. Similarly to maneuver load induced load factors it is seen from Eq. 7.13 that altitude, velocity, and the physical characteristics of the aircraft determine the gust load factors. At higher altitudes the load factor decreases rapidly as the density is decreased. This effect is offset somewhat due to higher airspeeds required to sustain flight at high altitudes. But, as already pointed out, gusts occur very infrequently at high altitudes.

The effect of airspeed on gust load factor is important for flight performance and operations. Since the gust load factor increases linearly with the velocity the effect is more pronounced at higher speeds. Thus, an aircraft operating at high speeds is expected to avoid turbulent regions or to reduce airspeed. Otherwise, structural damage may result. For the example aircraft in Figure 7.2, the gust load factors can be determined from Eq. 7.14. One obtains

$$n = 1 + \frac{0.002377 \times 0.08 \times 57.3 \times 30 \times 1.467 \times V}{2 \times 20{,}000/650}$$

$$= 1 + 0.00779V \quad \text{V mph, positive lift, sea level}$$

$$= 1 - 0.00779V \quad \text{V mph, negative lift, sea level}$$

The results are shown in Fig. 7.2 as 30 fps gust lines. For this particular aircraft at sea level, the gust produced load factors are expected to be less than the limit load factor of 6. The same aircraft with an improved powerplant and an increased $V_{max} = 750$ mph should not be operated at a speed higher than 642 mph when turbulence is anticipated. Higher speeds will produce a gust load factor higher than 6, which points to a probability of structural damage. At 20,000 ft altitude ($\sigma = 0.5328$), this aircraft can increase its speed to 1,204 mph and still stay within safe load-factor limits.

The lift curve slope, a, is another factor that indicates aircraft response to gust loads. A low-aspect-ratio, swept-wing aircraft has a low lift curve slope and will experience lower gust load factors. An aircraft with a high lift curve slope, and straight, high-aspect-ratio wings, would be more sensitive to gusts.

At the first glance it seems that an aircraft with higher wing loading, W/S, will be less sensitive to gusts. This is quite true when comparing two different aircraft, but can be somewhat misleading when one particular airplane is considered at two different weights. At low weights the gust-induced accelerations are higher, and at full loads the reverse is true. For given atmospheric conditions same lift acts on different masses and produces different accelerations as sensed by the pilot. The aircraft, however, is subjected to same loads.

The V–n diagram in Figure 7.2 shows the (safe) flight boundaries in level flight or in vertical plane maneuvers. The boundaries have been established from maximum lift capabilities or from structural strength limitations. Additional boundaries may be imposed by thrust limitations. The ceiling of an aircraft, which is thrust limited, has already been discussed in Chapter 3. Thrust-limited turns will be considered in the next section. From the point of view of aircraft operations, the lift-related boundaries are the most significant ones and will be considered again briefly below.

The lines of maximum lift capability indicate that an aircraft cannot produce higher steady-state load factors aerodynamically and is confined to fly below this line. This is true at both positive and negative lift coefficients except for the flight speeds at corresponding load factors, which are higher at negative lift coefficients due to a smaller negative maximum lift coefficient. At increasing flight speeds, even below the lift limit line, an aircraft can generate very large load factors that can exceed the limit load factor or even the ultimate load factor. If flight speeds are restricted to values below point A (intersection of the positive limit load factor and the line of maximum positive lift) then the aerodynamically generated load factors remain in the safe range below the limit load factor. At speeds greater than point A the aircraft is capable of producing positive lift and load factors that can cause damage to its structure. The same argument holds for negative lift and the corresponding point B.

The velocity corresponding to point A is the minimum airspeed at which the limit load factor can be produced aerodynamically. It is called *maneuver speed* or the *corner speed,* as it will give minimum turn radius and is the maximum velocity for alleviating stall due to gust. Combined aerodynamic and predictable gust loads cannot damage aircraft structure if the aircraft is flying below the corner speed. Thus, the corner speed is a useful reference point since below this speed one does not anticipate damaging flight loads. At speeds higher than the corner speed careful attention must be paid to the airspeed and the g load to be attempted. The corner speed is evaluated from

$$V_c = V_s\sqrt{n_1} \tag{7.15}$$

and fully reflects an aircraft's physical properties, aerodynamic lifting capability, and the altitude at which it is flying.

Corsair and Stuka

7.4 TURNING FLIGHT IN HORIZONTAL PLANE

For normal turning flight the aircraft is rolled to produce an unbalance in static equilibrium by rotating the lift vector out of the vertical plane.

The resulting force unbalance and acceleration then cause the aircraft to turn. A turn is primarily described by turn rate and turn radius, which are controlled by aircraft lifting, structural and thrust characteristics.

In a steady, coordinated turn the inclined lift vector produces a horizontal force component to equal the centrifugal force and a vertical component to balance the weight of the aircraft, Figure 7.4; $W = L \cos \phi$.

The bank angle ϕ is such that the above forces are exactly balanced and no sideways motion (yaw, sideslip) occurs. If a side force is generated, mainly for path correction, it can cause lateral acceleration leading to a turn at zero bank angle. This is called a *flat turn.* For a coordinated level turn, a simple force balance permits evaluation of the load factor from Figure 7.4 as

$$n \equiv \frac{L}{W} = \frac{1}{\cos \phi} \tag{7.16}$$

which shows that each bank angle produces a specific load factor n. For example, a bank angle of 60° requires a load factor of 2 for a steady coordinated turn and therefore a corresponding lift of $L = 2W$. If a different lift is generated at the same bank angle (through change of T and V), then acceleration components would exist (in both horizontal and vertical planes) and the turn would not be steady. As already pointed out, there should not be any aerodynamic forces in the spanwise direction and the aircraft should be pointed along the velocity vector. Strictly speaking, no yaw is allowed but the aircraft may be pitched in the lift vector plane (plane of aircraft symmetry).

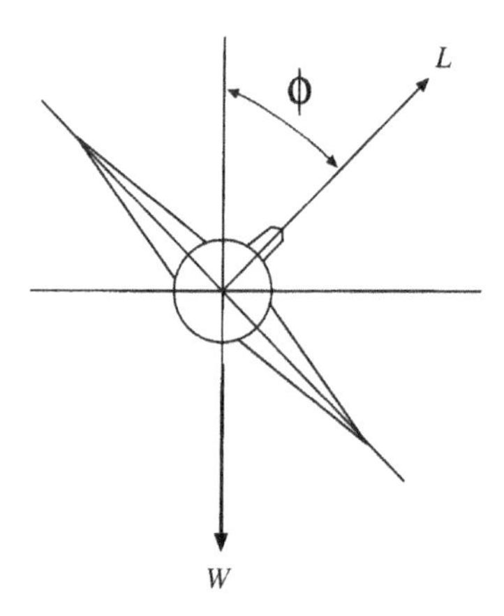

Figure 7.4 Turning Flight

To obtain expressions for turning velocity, radius, bank angle, and time, and to better define the coordinated turn, consider the trajectory and the forces acting on an aircraft when turning in a horizontal plane, Figure 7.5.

As shown in the Figure 7.5, T, L, D, and V are all acting in the plane of symmetry which is inclined to the vertical at the bank angle ϕ. The thrust T may be inclined at angle ϵ with respect to the horizontal plane. Weight W is in the vertical plane. While the aircraft is turning the flight path and the velocity vector remain in the horizontal plane and plane of symmetry and are oriented at angle ϕ from a (reference) vertical plane.

The governing equations are easily obtained by referring to Figure 2.1 and to the horizontal plane and plane of symmetry in Figure 7.5.

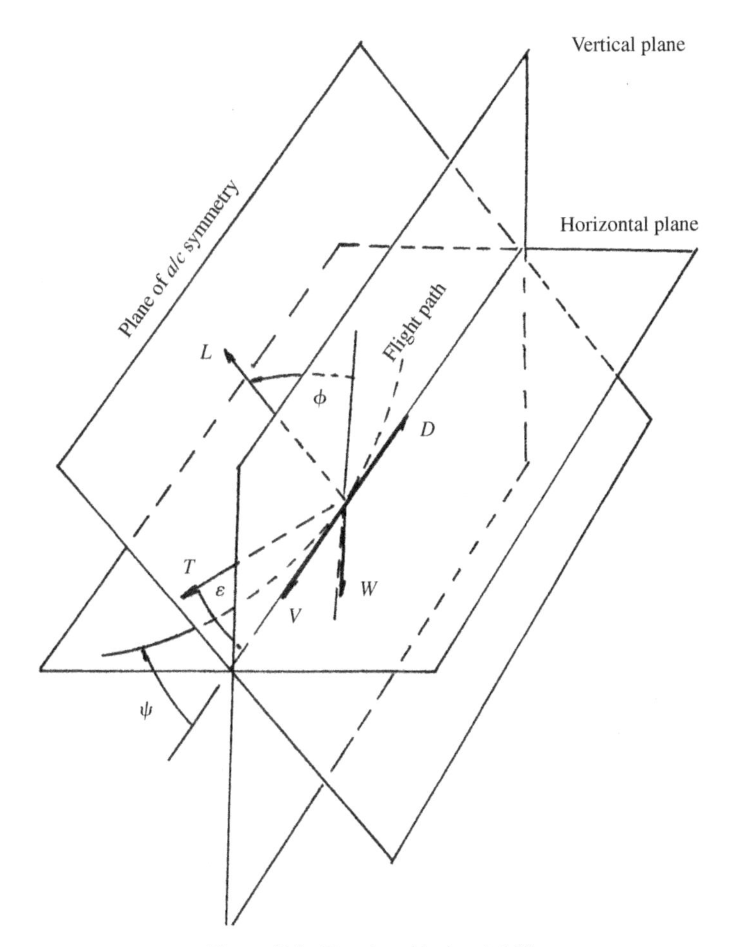

Figure 7.5 Turn in a Horizontal Plane

The equation tangent to the flight path is

$$T \cos \epsilon - D = m\dot{V}, \tag{7.17}$$

In horizontal plane,

$$(T \sin \epsilon + L) \sin \phi = mV\dot{\psi} = m \frac{V^2}{R} \tag{7.18}$$

In the vertical plane,

$$(T \sin \epsilon + L) \cos \phi = mg \tag{7.19}$$

Some basic results can now be obtained without further assumptions. Dividing Eq. 7.18 by Eq. 7.19 one gets

$$\tan \phi = \frac{V^2}{gR} \tag{7.20}$$

which shows that the bank angle is proportional to the velocity squared and inversely proportional to the radius of turn. The turn rate may be expressed as

$$\dot{\psi} = \frac{V}{R} \tag{7.21}$$

which becomes with Eq. 7.20, upon eliminating R:

$$\dot{\psi} = \frac{g}{V} \tan \phi \tag{7.22}$$

Since the bank angle was obtained as a function of the load factor, the turn rate and radius of turn may also be expressed as a function of the load factor. Thus, from

$$n = \frac{1}{\cos \phi}$$

and

$$\sin \phi = \sqrt{1 - \cos^2 \phi} = \sqrt{1 - \frac{1}{n^2}}$$

one obtains

$$\tan \phi = n\sqrt{1 - \frac{1}{n^2}} = \sqrt{n^2 - 1}$$

Then, Eqs. 7.20 and 7.22 become

$$R = \frac{V^2}{g \tan \phi} = \frac{V^2}{g\sqrt{n^2 - 1}} \tag{7.23}$$

and

$$\dot{\psi} = \frac{g}{V}\sqrt{n^2 - 1} \tag{7.24}$$

To estimate the time to turn through an angle ψ, it will be assumed that the flight path is a circle which then yields for time to turn through ψ radians

$$t = \frac{2\pi R}{V}\frac{\psi}{2\pi} = \frac{\psi R}{V} \text{ (sec)} \tag{7.25}$$

Eliminating now R by means of Eq. 7.23, one finds

$$t = \frac{V\psi}{g \tan \phi} \tag{7.26}$$

For unaccelerated level flight $\dot{V} = 0$, and Eqs. 7.17 and 7.19 can be combined to give the turning velocity

$$V = \sqrt{\frac{2Wn}{\rho S}\frac{1}{C_L + C_D \tan \epsilon}} \tag{7.27}$$

Combining Eqs. 7.26 and 7.27 gives

$$t = \frac{\psi}{g}\sqrt{\frac{2W}{\rho S\,(C_L + C_D \tan \epsilon)}}\sqrt{\frac{n}{n^2 - 1}} \tag{7.28}$$

and the turn radius becomes

$$R = \frac{2Wn}{g\rho S\,(C_L + C_D \tan \epsilon)}\sqrt{n^2 - 1} \tag{7.29}$$

Furthermore, the rate of turn can now be written

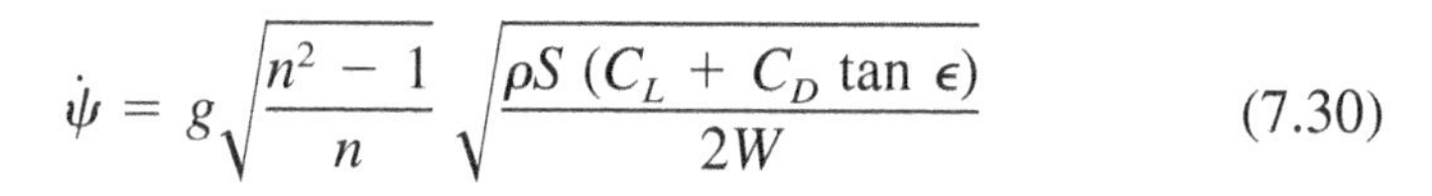

$$\dot{\psi} = g\sqrt{\frac{n^2 - 1}{n}}\sqrt{\frac{\rho S\,(C_L + C_D \tan \epsilon)}{2W}} \tag{7.30}$$

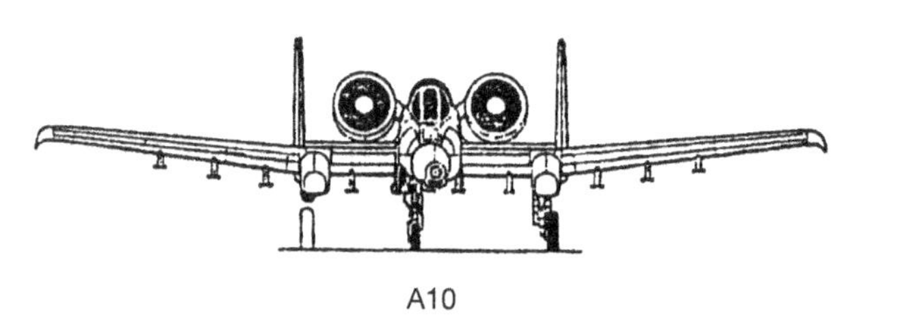

A10

A number of conclusions can be drawn concerning R, t, and $\dot{\psi}$. First of all, a practical assumption needs to be made concerning the load factor n (or the bank angle ϕ). Since a bank angle 90° implies a load factor of infinity, and since from previous discussions concerning the V–n diagram an aircraft is structurally limited to a specified value of $n = n_1$ (see Figure 7.2), both the bank angle and the load factor are then limited by the limit load factor n_1. However, a more serious limitation may be imposed by the available thrust, as is seen shortly.

From Eqs. 7.23 and 7.29, it is seen that, for a given maximum n_1, the minimum turn radius is obtained at lowest airspeed or at maximum lift coefficient $C_{L_{max}}$. As altitude increases, turn radius also increases due to decreased density. It follows, then, that the shortest turn is made when the aircraft is on the point of stalling. The same conclusions apply to the fastest turn (minimum time turn) when one observes Eqs. 7.26 and 7.28. Then $C_{L_{max}}$ is the desired condition for the maximum turn rate $\dot{\psi}$. This follows easily from Eq. 7.30 or by considering that $\dot{\psi}$ is inversely proportional to R for which minimum conditions were being considered.

Therefore, it seems that the best turning performance is governed by aerodynamic and structural limitations and that minimum turn radius and maximum turn rate are obtained at a point where $C_{L_{max}}$ and maximum load factor n_1 are obtained at the same time. This can also be seen from Figure 7.6, where lines of constant $\dot{\psi}$ and R are superimposed on the V–n diagram. This condition was already established in Section 7.3 as the corner velocity (maneuver speed), which is defined also by Eq. 7.27:

$$V_c = \sqrt{\frac{2Wn_1}{\rho S\,(C_{L_{max}} + C_D \tan \epsilon)}} = V_s\sqrt{n_1} \tag{7.31}$$

and then also a relationship between stall-in-turn and stall velocities V_{s_t} and V_s, respectively

$$V_{s_t} = V_s\sqrt{n}$$

Thus, in absence of any other restrictions, one obtains

$$R_{min} = \frac{V_c^2}{g\sqrt{n_1^2 - 1}} \tag{7.32}$$

$$\dot{\psi}_{max} = \frac{g\sqrt{n_1^2 - 1}}{V_c} \tag{7.33}$$

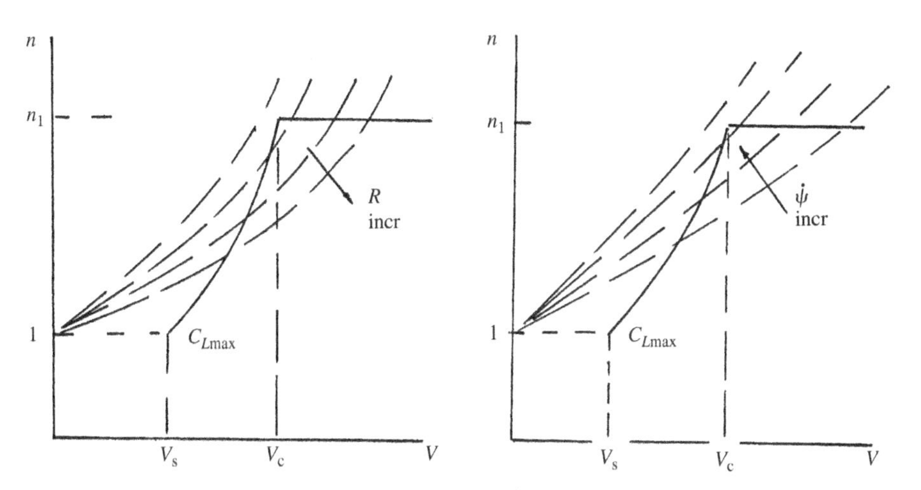

Figure 7.6 Constant R and $\dot{\psi}$ on V–n Diagram

$$t_{\min} = \frac{\psi}{g} V_c \sqrt{\frac{n_1}{n_1^2 - 1}} \tag{7.34}$$

$$V_{\dot{\psi}\max} = V_{R_{\min}} = V_c \tag{7.35}$$

It should be noticed that Eqs. 7.21, 7.23, and 7.24 are generic kinematic expressions, whereas Eqs. 7.32 and 7.33 are aircraft-specific via V_s ($C_{L\max}$) and the limit load factor n_1.

Summarizing: Aircraft turning performance, as governed only by aerodynamic and structural limitations, often referred to as instantaneous performance, consists of turns made at an instant of time without any consideration of sustaining that performance for any period of time. Of all the theoretical values of $\dot{\psi}$ and R, only the values inside the maneuver bounds (defined by $C_{L\max}$, n_1 curves, and $V_{\max}$, see Figure 7.2) are available. At speeds below the corner speed V_c turning performance is aerodynamically limited by $C_{L\max}$. Above V_c, maximum aerodynamic load factor can be larger than n_1 and then the turning performance is structurally limited. Eqs. 7.32 and 7.33 do not assure that such maxima can be achieved. Rather, they establish potential limits subject to other constraints (thrust). It is also evident that the (instantaneous) turn rate and radius are basically determined, through V_c, by the maximum lift coefficient $C_{L\max}$ and wing loading W/S. These two parameters play a significant role in sustained maneuvering performance (see also Problem 7.11).

In practice, a much more severe restriction is placed on the turning performance by the fact that the thrust, or power, is limited. For high-performance aircraft, where $T/W \approx 1$, this will not be a problem at low altitudes. But degradation of engine thrust at high altitudes will impose turning restrictions on all aircraft.

As a result of lift increase required to produce a turn while banking, the induced drag is increased above the level flight value. This drag increase can be observed, for steady flight, from Eqs. 7.17 and 3.22 as

$$T_r = D = \frac{1}{2}\rho V^2 S C_{D0} + \frac{2kW^2}{\rho V^2 S \cos^2 \phi} \tag{7.36}$$

When Eq. 7.36 is plotted for constant ρ and W, as a function of velocity and bank angle with the thrust available curves also present, the increase in induced drag and the role of the bank angle are clearly evident. A bank angle of $\phi = 0$ represents the steady, level flight results. Figure 7.7 shows that as the bank angle (load factor) increases,

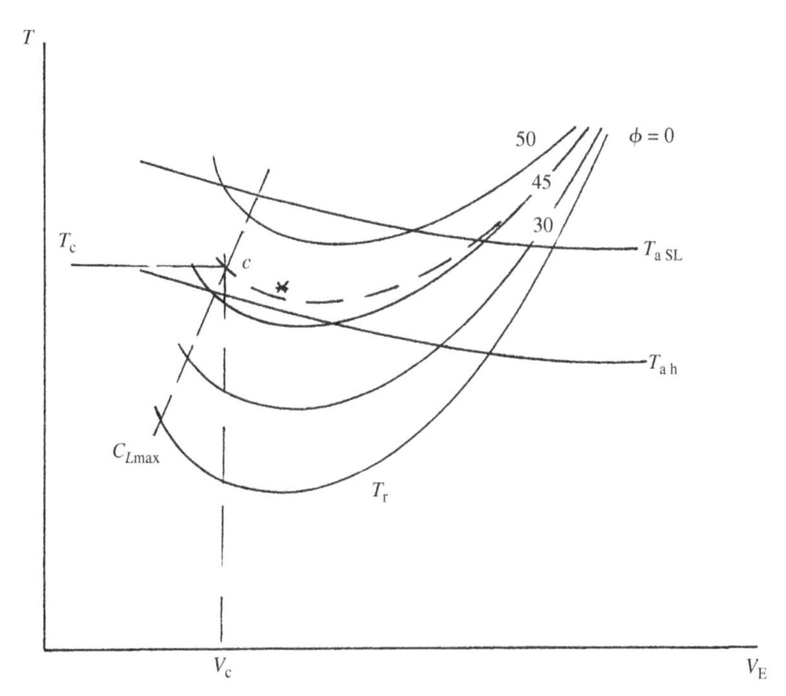

Figure 7.7 Effect of Bank Angle on Thrust Required

the drag may exceed the thrust available. To maintain constant altitude, the bank angle must be decreased to the point where $T = D$. Thus, the turning performance becomes thrust limited. Turns at higher bank angle (higher load factor) may be accomplished, but the altitude and velocity cannot be maintained at the same time.

Thrust limitation also influences the minimum radius and maximum turn rate performance. Figure 7.8 shows constant thrust curves superimposed on the V–n diagram. The significance of these curves lies in the fact that they show the amount of thrust required to maintain corner velocity. This, of course, can be calculated by use of Eqs. 7.31 and 7.36. If there is sufficient thrust available to maintain V_c ($T_a \geq T_r$), the T_r curve must pass above V_c–n_1 intersection and the maximum turning performance Eqs. 7.32 through 7.35 hold.

For the configuration shown in Figure 7.7 one can conclude that at T_a shown for the sea level turning performance is not thrust limited. At altitude h, however, T_{a_h} curve falls below T_c point, as required by V_c, and there will be an increase in R_{min} and a reduction of $\dot{\psi}_{max}$ from the values predicted by Eqs. 7.32 and 7.33, respectively. Those equations, or rather Eqs. 7.23 and 7.24, can still be used to evaluate R and

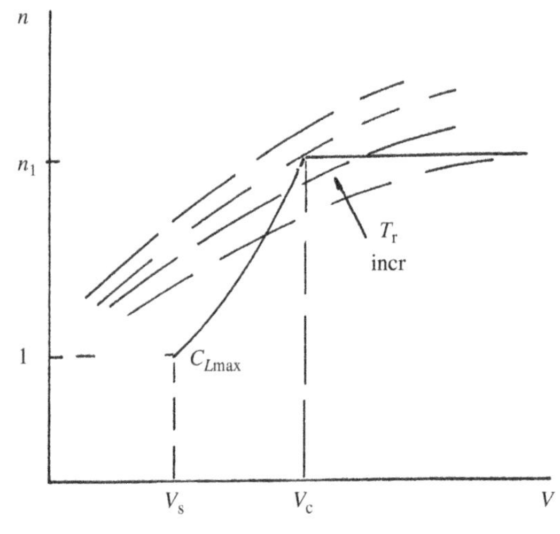

Figure 7.8 Constant T_r Lines

$\dot{\psi}$, but a new load factor must be determined from Eq. 3.21 (or any similar expression) by use of the actual thrust available at that particular altitude.

At thrust levels where $T_{ah} < T_c$, turns can be made at $n \leq n_l$, the structural limit, but altitude and/or velocity may change (decrease), allowing the recovered energy to be converted into a temporary turn. What actually happens is governed by the energy equation Eq. 2.20. A number of options are available, depending on operational needs. Such problems fall in the area of energy maneuverability, some of which can be approached via the point-performance techniques and will be discussed in Section 7.6.

7.5 MAXIMUM SUSTAINED TURNING PERFORMANCE

In the previous section, aircraft maneuvering performance was found to be constrained primarily by its aerodynamic and structural characteristics leading to expressions describing the instantaneous performance potential. It was found also that the thrust was the deciding factor whether this potential can be achieved and then maintained.

In practical flight-performance analysis, the items of great interest are the maximum sustained turn rate and the minimum sustained turn radius. Implicit in such a development is the assumption that the motion and aircraft attitude be steady and could be maintained for some practical, but undetermined, time. Returning to Eq. 2.20 where aircraft performance was formulated in terms of the energy balance, it is seen that *steady* and *maintained* would have to mean that the altitude and velocity must remain constant and therefore the energy rate must be identically zero ($P_s = 0$, see Eq. 3.1). It follows also that $T = D$, and the problem falls clearly into point-performance category. This is called sustained performance. Its meaning and definition have found widespread use in aircraft performance description, comparison, and even acquisition.

7.5.1 Maximum Load Factor

First of all, it is helpful to develop more understanding and an expression for the maximum load factor. This is obtained by rewriting the thrust required ($T = D$) expression Eq. 3.21 as:

$$n^2 = \left[\frac{q}{kW/S}\left(\frac{T}{W} - q\frac{C_{D0}}{(W/S)}\right)\right] \tag{7.37}$$

or

$$n = \frac{q}{W/S}\sqrt{\frac{1}{k}\left(\frac{T/S}{q} - C_{D0}\right)} \tag{7.38}$$

Eqs. 7.37 and 7.38 give the load factor for a given thrust loading T/W and velocity ($q = 1/2\rho V^2$). For each value of q the local maximum value of the load factor is determined by the maximum value of T/W (see Figure 7.8). The velocity that maximizes the load factor is found from Eq. 7.38 by differentiation with respect to q and then setting the result equal to zero. One obtains for the dynamic pressure at maximum load factor

$$q_{\text{nm}} = \frac{T/S}{2C_{D0}} \tag{7.39}$$

which yields for the velocity

$$V_{\text{nm}} = \sqrt{\frac{T/S}{\rho C_{D0}}} \tag{7.40}$$

Substituting now Eq. 7.39 into Eq. 7.37 gives

$$n_{\max} = \frac{T}{W} E_{\mathrm{m}} \tag{7.41}$$

Eq. 7.41 shows that the maximum load factor is obtained at maximum lift-drag ratio, E_{m} *at a given T/W ratio* (any T/W). The true maximum available load factor is obtained also at E_{m} but at *the maximum available* thrust-weight ratio $(T/W)_{\max}$. In other words, this occurs when the aircraft is operating simultaneously at both maxima:

$$n_{\mathrm{m}} = \left(\frac{T}{W}\right)_{\mathrm{m}} E_{\mathrm{m}} \tag{7.42}$$

The corresponding flight speed can be brought in sharper focus if Eq. 7.40 is nondimensionalized by $V_{D_{\min}}$, obtaining

$$\overline{V}_{n_{\mathrm{m}}} = \frac{V_{n_{\mathrm{m}}}}{V_{D_{\min}}} = \sqrt{\frac{T/S}{\rho C_{D_0}}}\sqrt{\frac{\rho}{2W/S}}\sqrt{\frac{C_{D_0}}{k}} = \sqrt{\frac{T}{W}E_{\mathrm{m}}} \tag{7.43}$$

Eqs. 7.40 and 7.43 show that the velocity for maximum available load factor is proportional to the available thrust. It decreases as the altitude increases with the decrease of thrust with altitude.

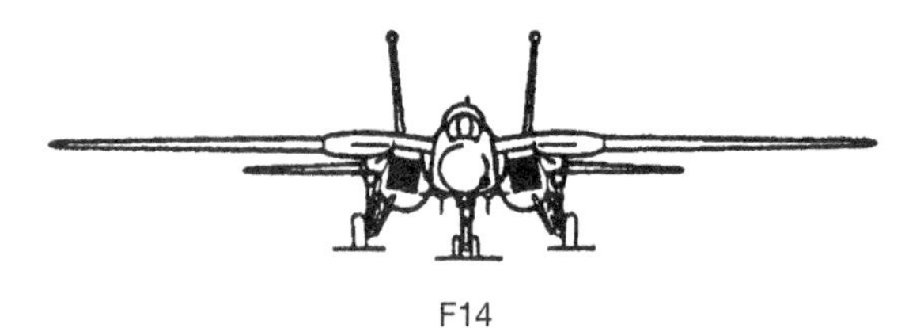

F14

7.5.2 Minimum Turn Radius

The minimum turn radius is found by differentiating Eq. 7.23:

$$R = \frac{V^2}{g\sqrt{n^2 - 1}}$$

with respect to V and setting the result equal to zero. Since dn/dV is equal to $\rho V\, dn/dq$, it is convenient to change variables, and one finds the well-known result

$$n^2 - 1 - qn\frac{dn}{dq} = 0 \tag{7.44}$$

Substituting now dn/dq from Eq. 7.37, after some manipulation, one obtains the following results for the condition of minimum turn: velocity of the turn (assumed to be constant):

$$V_{R_{\min}} = \sqrt{\frac{4k(W/S)}{\rho(T/W)}}, \tag{7.45}$$

applicable load factor, from Eqs. 7.37 and 7.45:

$$n_{R_{\min}} = \sqrt{2 - \frac{1}{\left(\frac{T}{W}\right)^2 E_m{}^2}}, \tag{7.46}$$

minimum turn radius:

$$R_{\min} = \frac{V^2_{r_{\min}}}{g\sqrt{n_{r_{\min}}{}^2 - 1}}, \tag{7.47}$$

the turning rate for minimum radius turn follow form Eq. 7.33:

$$\dot{\psi}_{R_{\min}} = \frac{g\sqrt{n^2_{R_{\min}} - 1}}{V_{R_{\min}}}, \tag{7.48}$$

or

$$\dot{\psi}_{R_{\min}} = \frac{g \tan \phi_{R_{\min}}}{V_{R_{\min}}}, \tag{7.49}$$

where

$$\phi_{R_{\min}} = \frac{1}{n_{R_{\min}}}. \tag{7.50}$$

Eqs. 7.45 to 7.50 provide necessary tools for evaluating the minimum turn radius and, at the first glance, show that high thrust-weight ratio and a high E_m with a low wing loading provide the desired results. However, a comparison of the $V_{R_{min}}$ with the stall velocity in shows that if the available thrust-weight ratio T/W is higher than $2kC_{L_{max}}/n_{R_{min}}$, obtained by equating V_s

$$V_s\sqrt{n} = V_{R_{min}} \tag{7.51}$$

in Eq. 7.45, then $V_{R_{min}} < V_s$ and the turn cannot be accomplished aerodynamically. In other words, the locus of $(V_{R_{min}}, n)$ is on the T/W curve but to the left of the stall line. In this case, the minimum turn radius, for that given T/W, is on the stall line where the T/W curve crosses the stall line. That intersection can be found by equating the load factors determined from the equation of the stall line:

$$n = \frac{q}{q_s} = \frac{qC_{L_{max}}}{(W/S)} \tag{7.52}$$

to the load factor from Eq. 7.37:

$$\frac{qC_{L_m}}{W/S} = \left(\frac{q}{k(W/S)}\left[\frac{T}{W} - \frac{C_{D_0}q}{W/S}\right]\right)^{1/2} \tag{7.53}$$

This gives the stall speed in turn as

$$q_{R_{min}} = \frac{(T/W)(W/S)}{C_{D_{min}}} \tag{7.54}$$

where $C_{D_m} \equiv kC_{L_m}{}^2 + C_{D_0}$.

The corresponding stall load factor is obtained by substituting Eq. 7.54 back into Eq. 7.52, giving

$$n_{r_{min}} = \frac{(T/W)C_{L_{max}}}{C_{D_m}} \equiv \frac{T}{W}E_2 \tag{7.55}$$

where

$$E_2 \equiv \frac{C_{L_{max}}}{C_{D_{max}}} \tag{7.56}$$

Thus, a turn can be made according to Eqs. 7.54 and 7.55 but at higher turning radius, which can then be calculated again from Eq. 7.47 but with values of V and n at $C_{L_{max}}$, or, the turn is made at edge of stall.

NOTE

Since the stall line velocity, determined from Eq. 7.53, must be equal (on the stall line) to the velocity obtained from Eq. 7.45, a simpler check for the turn can be found by equating these equations. Thus, one finds that if

$$2kC_{D_m} \geq \left(\frac{T}{W}\right)^2$$

the turn can be made. Then $V_{R_{min}} \leq V_{R{min}_s}$ and the locus of (V_{Rmin}, n) is on the T/W curve but on, or to the right of, the stall line.

7.5.3 Maximum Turning Rate

The maximum steady-state turning rate, also called the fastest turn, is found through a process similar to that used in the previous section for the minimum turn radius. The equation for the turn rate, Eq. 7.33, is differentiated with respect to q. Setting the result equal to zero gives the basic condition for the maximum turning rate as

$$n^2 - 1 - 2nq\frac{dn}{dq} = 0 \tag{7.57}$$

When combining with Eq. 7.38, one obtains for the fastest turning velocity

$$q_t = \frac{W}{S}\sqrt{\frac{k}{C_{D0}}} \tag{7.58}$$

whence

$$V_t = \sqrt{\frac{2(W/S)}{\rho}\left(\frac{k}{C_{D0}}\right)^{.25}} = V_{D_{min}} \tag{7.59}$$

is the velocity in level flight at that particular altitude. Recombining Eq. 7.58 with Eq. 7.38 gives

$$n_{\mathrm{t}} = \sqrt{2\frac{T}{W}E_{\mathrm{m}} - 1}$$

and

$$n_{\mathrm{t_{max}}} = \sqrt{2\left(\frac{T}{W}\right)_{\mathrm{m}} E_{\mathrm{m}} - 1} = \sqrt{2n_{\mathrm{max}} - 1}. \tag{7.60}$$

Recalling that $L = nW$, Eq. 7.58 also yields a limiting condition

$$C_{L_{\mathrm{t}}} = n_{\mathrm{t}}\sqrt{\frac{C_{D0}}{k}} = n_{\mathrm{t}}\, C_L|_{EM} \le C_{L\,\mathrm{max}} \tag{7.61}$$

The maximum turning rate can be calculated simply from Eq. 7.33 as

$$\dot{\psi} = \frac{g}{V_{\mathrm{t}}}\sqrt{n_{\mathrm{t}}^2 - 1} \tag{7.62}$$

Using Eq. 7.58 again, one obtains a more explicit equation:

$$\dot{\psi} = g\left(\frac{\rho}{2(W/S)k}\left[\frac{T}{W} - 2\sqrt{kC_{D0}}\right]\right)^{1/2} \tag{7.63}$$

It is seen that a high turning rate requires a large T/W but low values of the wing loading W/S, k, and C_{D0}. Small k means large aspect ratio AR. Small k and C_{D0} imply also a high E_{m}. Thus, conflicts and necessary compromises arise in the design stage.

Transport and small noncombatant aircraft rely on high E_{m} and relatively low T/W values for economic operation. High turning rates become significant only for acrobatic categories. Fighter aircraft need low aspect ratio for high speed and structural considerations. Thus, high T/W and $C_{L\,\mathrm{max}}$ are then usual trade-offs for lower aspect ratio and E_{m} with attendant range penalties. Eqs. 7.59 and 7.63 also show why air-to-air combat takes place at subsonic speeds ($V_{D\mathrm{min}}$) and on deck due to maximum air density at the sea level.

Eqs. 7.60 and 7.61 highlight another hidden constraint. A high-performance aircraft with a high T/W may be able to produce theoretically a sufficiently high n_{max} to generate, in turn, high lift requirement ($C_{L_{\mathrm{t}}}$ in Eq. 7.61), which may exceed the aircraft maximum lifting capability $C_{L_{max}}$. Then, similarly to minimum turn radius case, the turn may have to be performed (at reduced T/W) at lift limit curve at (near) stall conditions.

In summary, all three limitations due to aerodynamics, structural considerations, or thrust may affect at the same time aircraft turning performance. In general, aerodynamic and structural limitations are the significant ones at low altitudes. At high altitudes, thrust considerations predominate. At high airspeeds $C_{L_{max}}$ may also be limited due to Mach number effects.

EXAMPLE 7.1

An aircraft has the following characteristics:

$C_{L_{max}} = 1.5$
$S = 125 \text{ ft}^2$
$W = 2{,}000 \text{ lb}$
$n_1 = 4$
$C_D = 0.02 + 0.05C_L^2$
$T = 300$ lb for all flight speed at sea level (turbojet)
$\epsilon = 0$

Calculate the minimum time added to perform a constant altitude 180° turn. Assuming that there will be adequate thrust to perform the turn, Eqs. 7.31 and 7.34 can be used. The drag coefficient is

$$C_D = 0.02 + 0.05(1.5)^2 = 0.1325$$

and the stall speed is

$$\begin{aligned} V_s &= \sqrt{\frac{2W}{\rho S C_{L_{max}}}} \\ &= \sqrt{\frac{2 \times 2{,}000}{0.002377 \times 125 \times 1.5}} \\ &= 94.7 \text{ ft/sec} \end{aligned}$$

The corner speed becomes

$$V_c = V_s\sqrt{n_l} = 95\sqrt{4} = 190 \frac{\text{ft}}{\text{sec}}$$

The time for minimum turn follows from

$$t_{\min} = \frac{\psi V_c}{g\sqrt{n_l^2 - 1}} = \frac{\pi \times 190}{32.2 \times \sqrt{15}} = 4.79 \text{ sec}$$

Checking now the validity of adequate thrust assumption

$$T_r = \frac{1}{2}\rho V^2 S C_D = \frac{1}{2} \times 0.002377 \times (190)^2 \times 125 \times 0.1325 = 710 \text{ lb}$$

which clearly exceeds the thrust available and indicates that a 4g turn at the corner speed is not possible. What is actually possible can be calculated as follows: the load factor at the available thrust of 300 lb is evaluated from

$$n_t = \sqrt{2\left(\frac{T}{W}\right)E_m - 1} = \sqrt{2 \times \frac{300}{2{,}000} - 1} = 1.934$$

In order to calculate the flight velocity, the turn lift coefficient is obtained from Eq. 7.61

$$C_{L_t} = n_t\sqrt{\frac{C_{D0}}{k}} = 1.934\sqrt{\frac{.02}{.05}} = 1.22$$

Turn velocity is

$$V_t = \sqrt{\frac{2Wn_t}{\rho S C_{L_t}}} = \sqrt{\frac{2 \times 2000 \times 1.934}{.002377(125)1.22}} = 146 \text{ ft/sec}$$

Bank angle is calculated from

$$\tan\phi = \sqrt{(n_t)^2 - 1} = \sqrt{1.934^2 - 1} = 1.655$$

which indicated a bank angle of 59°. The time to turn is obtained from Eq. 7.26

$$t = \frac{V\psi}{g \tan \phi} = \frac{146\pi}{32.2 \times 1.655} = 8.59 \text{ sec}$$

This problem can be solved also in an indirect manner from Eqs. 7.27 and 7.28, which can be written as

$$\frac{tg}{\psi} = \frac{V}{\sqrt{n^2 - 1}}$$

The problem now resolves into one of finding the combination of V and n that minimizes tg/ψ. This can be determined by solving the $T_a = D$ equation for n, for selected values of V, and then calculating $V/\sqrt{n^2 - 1}$. The drag equation becomes

$$T_a = \frac{1}{2} \rho V^2 S C_{D_0} + \frac{2kW^2n^2}{\rho S V^2}$$

or

$$300 = \frac{1}{2} \times 0.002377 \times V^2 \times 125 \times 0.02 + \frac{2 \times 0.05 \times (2{,}000)^2 \times n^2}{0.002377 \times 125 \times V^2}$$

Which simplifies to

$$\frac{300V^2 - 0.002971V^4}{1346235} = n^2$$

Constructing Table 7.1 gives a minimum value:

$$\frac{tg}{\psi} = 88.07$$

which occurs at a flight velocity of 146 ft/sec. The time to turn can be calculated from

TABLE 7.1 Data for Example 7.1

V	n	$V/(n^2 - 1)$
140	1.876	88.19
145	1.926	88.09
146	1.936	88.07
147	1.946	88.08
150	1.974	88.13

$$t = \frac{\pi \times 88.07}{32.2} = 8.59 \text{ sec}$$

The turn radius is calculated from

$$R = \frac{V^2}{g\sqrt{n^2 - 1}} = \frac{146^2}{32.2 \times \sqrt{1.936^2 - 1}} = 399 \text{ ft}$$

This process amounts to assuming that the lift coefficient C_L remains constant throughout the turn. In other words, in level flight prior to turning or when returning to level flight after the turn, the lift coefficient is the same as in the turn:

$$C_L = \frac{2Wn}{\rho SV^2} = \frac{2 \times 2000 \times 1.936}{0.002377 \times 125 \times (146)^2} = 1.22$$

Thus, when the aircraft returns to (or starts from) level flight, its flight speed is reduced to

$$V_L = \sqrt{\frac{2W}{\rho SC_L}} = \sqrt{\frac{2 \times 2000}{0.002377 \times 125 \times 1.22}} = 105 \frac{\text{ft}}{\text{sec}}$$

The same result can also be calculated from $V_L = V/\sqrt{n}$. Then the thrust for level flight can be reduced from 300 lb to [with $C_D = 0.02 + 0.05(1.22)^2 = 0.0944$]

$$T = \frac{1}{2} \times 0.002377 \times 0.0944 \times 125 \times (105)^2 = 155 \text{ lb}$$

EXAMPLE 7.2

The aircraft in Example 7.1 is in level flight at 125 ft/sec and sets out to perform the same 180° turn but keeps the flight velocity constant while turning. This, however, requires that the lift coefficient be increased to maintain altitude and to produce turning acceleration. Thus, the drag coefficient also will increase with an increase in thrust (over the level flight value of 155 lb). Eq. 7.26 permits calculation of the required load factor n from

$$t = \frac{V\psi}{g\sqrt{n^2 - 1}}$$

whence

$$n = \sqrt{\left(\frac{V\psi}{tg}\right)^2 + 1} = \sqrt{\left(\frac{125 \times \pi}{8.59 \times 32.2}\right)^2 + 1} = 1.74$$

The required C_L, and C_D, are obtained by means of Eq. 7.27, which gives

$$C_L = \frac{2Wn}{\rho SV^2} = \frac{2 \times 2{,}000 \times 1.74}{0.002377 \times 125 \times 125^2} = 1.5$$

and the drag coefficient as

$$C_D = 0.1325$$

The thrust required to accomplish this turn is

$$T = \frac{1}{2} \times 0.002377 \times 0.1325 \times 125 \times 125^2 = 307 \text{ lb}$$

and the resulting turn radius is

$$R = \frac{125^2}{32.2 \times \sqrt{1.74^2 - 1}} = 341 \text{ ft}$$

7.6 THE MANEUVERING DIAGRAM

The last two sections established the important parameters required for analyzing aircraft maneuvering performance in horizontal plane. Useful equations were obtained predicting minimum turn radius and the maximum turn rate. In addition, it was found that the maximum turn capability is limited in an intricate fashion by three factors:

1. Structural strength limit
2. Lifting capability
3. Thrust limit

All this information can be combined in a common presentation diagram called the *maneuvering diagram* or the *energy maneuverability diagram*. It has the advantage of presenting the aircraft turning capability in a very compact and complete manner. However, each altitude and gross weight require separate diagrams. The diagram is also very useful for presenting and comparing several tactical aircraft turning performances. As is seen in Figure 7.9, the horizontal axis represents the velocity and the vertical axis gives the turn rate. The basic graph consists of a carpet plot of the load factor and the turn radius. Figure 7.9, generated by Eqs. 7.21, 7.23, and 7.24, and by its purely kinematic content, is generic and valid for any aircraft. The aircraft-specific data is entered through the limiting factors.

The structural limit n_l and the lift limit (also called accelerated stall boundary) are entered via $V = V_s\sqrt{n_l}$ and superposed on the graph; see Figure 7.10. This introduces the aircraft-specific information through chosen altitude and weight data. Due to its shape, the resulting graph is also known as the doghouse plot. Figure 7.10 shows the curves of A-4 Skyhawk for the following data:

$$W = 18{,}000 \text{ lb}, \; C_D = .0177 + .156C_L^2, \; S = 260 \text{ ft}^2, \; T = 10{,}000 \text{ lb}$$

For illustration purposes it will be assumed that the limit load factor $n_l = 5$. Then from Fig. 7.10, at the point A, where the stall line crosses the locus of $n_l = 5$, one finds that

- The maximum (instantaneous) turn rate is about 21.5 degrees.
- The minimum (instantaneous) turn radius is about 1,200 ft.
- The corner speed is about 255 KTAS.

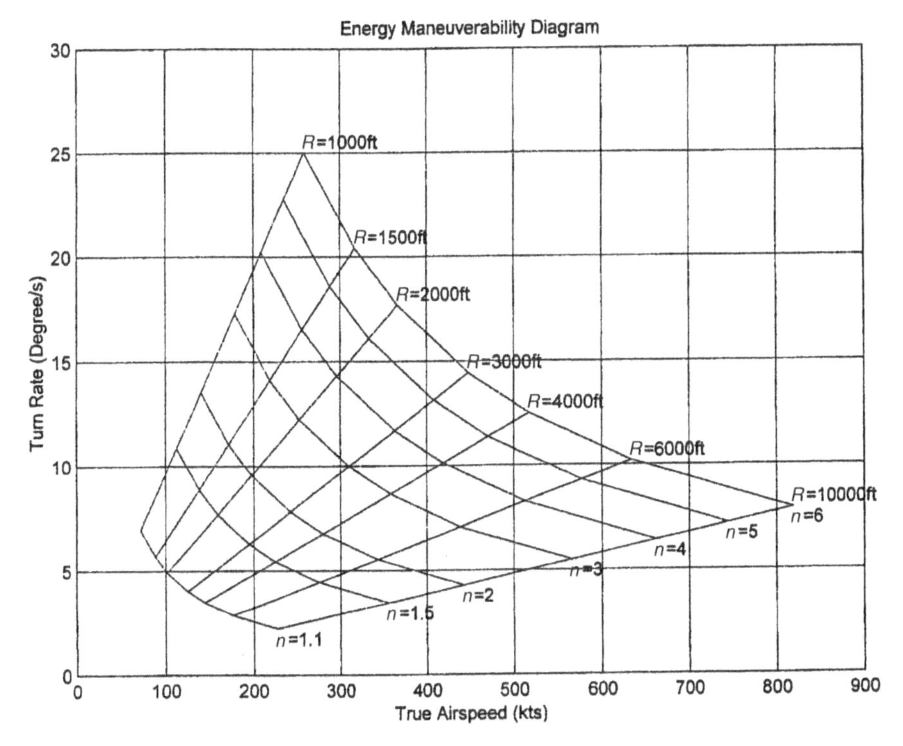

Figure 7.9 The Maneuvering Diagram

For evaluating sustained performance capabilities, it is necessary to consider also the thrust limit. This is accomplished via the energy equation (Eq. 2.20), which is used in the form

$$P_s \equiv \frac{de}{dt} = \frac{dh}{dt} + \frac{V}{g}\frac{dV}{dt} = \left(\frac{T - D}{w}\right) V$$

$$= Ma\left[\frac{T}{W} - \left(\frac{C_{D0}q}{WS} + kn^2\frac{W/S}{q}\right)\right]$$

where the drag has been introduced from Eq. 3.21. Next, a parametric curve calculation is established for an arbitrary range of $-500 \leq P_s \leq 250$ and a practical range of velocities with the load factor n as another parameter to assist in superimposing the P_s curves in Fig. 7.6. The curve for $P_s = 0$ signifies $T = D$, which provides the locus for sustained (steady) performance evaluation.

Following the line for load factor $n_l = 5$ to the $P_s = 0$ curve (point B) gives the following:

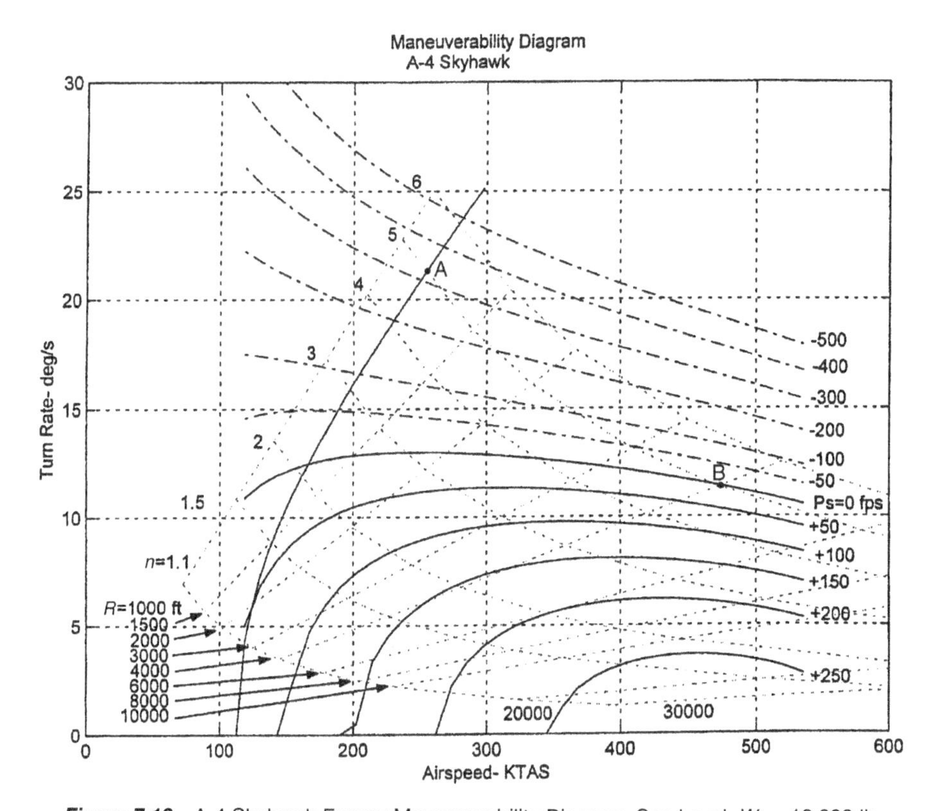

Figure 7.10 A-4 Skyhawk Energy Maneuverability Diagram, Sea Level, W = 18,000 lb

- The minimum sustained turn radius is little less than 4,000 ft at about 470 KTAS, with a turn rate of about 12 degrees (point B).
- The maximum sustained turn rate is found at the top of $P_s = 0$ curve at about 13 degrees/sec. This takes place at a load factor of little more than 3 and a turn radius of about 1,800 ft.

However, this is not the complete picture. To better understand the role of P_s it is useful to return to the energy rate equation Eq. 2.20 and briefly consider how energy can be gained and then redistributed in various maneuvers.

The energy production side of Eq. 2.20

$$P_s = \frac{T - D}{W} V$$

clearly shows that the aircraft specific energy is increased if $T > D$. Conversely, if $P_s = 0$ the aircraft has no capability to increase its specific energy and it represents the upper limit of sustainable maneuvering (at $T = T_{max}$). If P_s is positive, then the aircraft can (must) accelerate, climb, or maneuver. If P_s is negative, it can still perform some of these functions but now it also has to decelerate or lose altitude, as it is losing energy. Thus, the $P_s = 0$ curve is a useful boundary for separating sustained maneuvers from those that lead to energy loss.

For some specific examples, consider again the A-4 Skyhawk. At sea level 1-g loading and at 500 KTAS, its P_s is

$$P_s = \frac{(T - D)V}{W} = \frac{10{,}000 - 4{,}140}{18{,}000} 845 = 275 \text{ ft/sec}$$

From the energy conservation side of Eq. 2.20

$$P_s \equiv \frac{de}{dt} = \frac{dh}{dt} + \frac{V}{g}\frac{dV}{dt}$$

one can conclude that

- At constant KTAS climb ($dV/dt = 0$) the aircraft can climb at the rate of 275 ft/sec; or
- If the altitude h is constant ($dh/dt = 0$), the aircraft can accelerate at

$$\frac{dV}{dt} = \frac{275g}{V} = 275g/845 = 10.5 \text{ ft/sec}^2\text{; or}$$

- If the aircraft is set into a vertical climb where $dh/dt \equiv V = 275$ ft/sec, then

$$P_s = \frac{dh}{dt} + \frac{V}{g}\frac{dV}{dt} = V + \frac{V}{g}\frac{dV}{dt}$$

whence

$$\frac{dV}{dt} = \left(\frac{275}{845} - 1\right)g = -21.7 \text{ ft/sec}^2$$

and the aircraft is decelerating at about .67g.

Returning now to the example concerning instantaneous performance, if the pilot wants to perform a 5g turn at the corner speed (about

250 KTAS), the aircraft would lose altitude at the rate of about 330 ft/sec.

In its essence, the maneuvering diagram represents point-performance approach since it is tied to a given altitude and weight. Change of altitude has no effect on the limit load factor, as it is only a function of structural design. For a given load factor, altitude gain does increase the stall velocity. An increase in altitude will decrease the available thrust. The combined effect is to shrink the positive P_s region with attendant reduction of the turn rate and an increase of turning radius. Effect of the altitude change may be estimated by preparing the diagrams similar to the one in Fig. 7.10 at different altitudes and weight, but such procedure cannot really predict interaltitude behavior because of time dependent behavior of the aircraft velocity. Actual (time-dependent) problems are the topic of energy maneuverability applied to flight paths.

7.7 SPIRAL FLIGHT*

An aircraft that is rolled into a (steady) coordinated turn at a constant angle of bank ϕ, and flies at a constant angle or descent γ moves on a helical path on an imaginary circular cylinder with a radius of R, (see Figure 7.11).

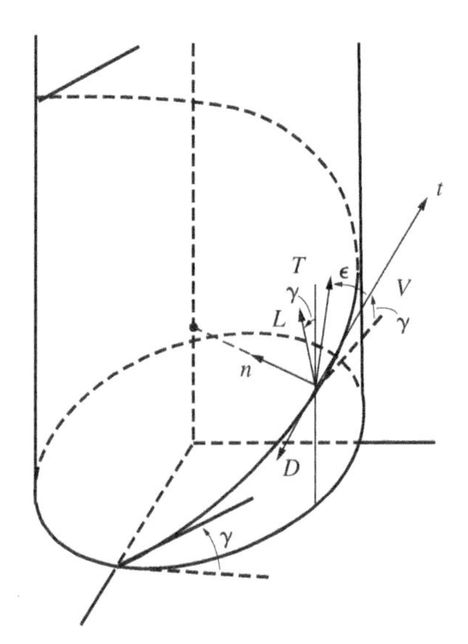

Figure 7.11 Spiral Flight

This spiral path can represent also instantaneous (point performance) conditions of a transient flight path where only centripetal and tangential accelerations are included. Thus, as it was assumed in Section 7.4, the forces are exactly balanced without any yaw or sideslip taking place. Referring to Figures 7.3 and 3.2, the equation tangent to the flight path is

$$T \cos \epsilon - D = W \sin \gamma + m\dot{V} \tag{7.64}$$

the equation normal to the flight path (along the radius R) is

$$(T \sin \epsilon + L) \frac{\sin \phi}{\cos \gamma} = mV\dot{\psi} = \frac{mV^2}{R} \cos \gamma \tag{7.65}$$

and in the vertical plane the equation is

$$(T \sin \epsilon + L) \cos \phi = W \cos \gamma \tag{7.66}$$

Here $\dot{\psi}$ is the turn rate about a vertical axis. The angle ϵ, in general, is not small for a hard banked turn, as the thrust sector may have a sizable inclination to the velocity sector. This is due to the thrust installation angle δ, and the aircraft angle of attack α, which will now approach the stall angle (usually exceeding 20°).

$$\epsilon = \delta + \alpha \tag{7.67}$$

In previous chapters the thrust angle ϵ was ignored by assuming either that the angle of attack was small or that the thrust contribution to the lifting effort was small (see Eq. 2.12). In a highly banked turn with required large amount of thrust, the angle ϵ will become important, as the thrust will contribute significantly to the turning lift force. Similar to the developments in Section 7.4, the load factor comes from Eq. 7.66 as

$$\frac{T \sin \epsilon + L}{W} = \frac{\cos \gamma}{\cos \phi} = n \tag{7.68}$$

Also

$$\tan \phi = \frac{V^2 \cos \gamma}{gR} \tag{7.69}$$

and

$$\dot{\psi} = \frac{V \cos \gamma}{R} = \frac{g \tan \phi}{V} \tag{7.70}$$

Eqs. 7.68 through 7.70 are equivalent to Eqs. 7.20 through 7.24 of section 7.4 and are valid for climbing or descending flight ($\gamma < 0$). For the case of flight without power (thrust), this set of equations can be developed further to yield some practical and interesting results. In earlier developments and examples it was shown that more power is needed in turns than in level flight. It will be shown now that as an airplane is descending in glide, the angle of glide will be increased if the plane is turning. In glide, as the plane turns through a heading of ϕ, the height lost is

$$\Delta h_{\psi} = \psi R \tan \gamma = \psi \frac{WV^2 \cos^2 \gamma}{gL \sin \phi} \tan \gamma \tag{7.71}$$

This can be simplified by

$$V^2 = \frac{2W/S}{\rho C_L \cos \phi}$$

and

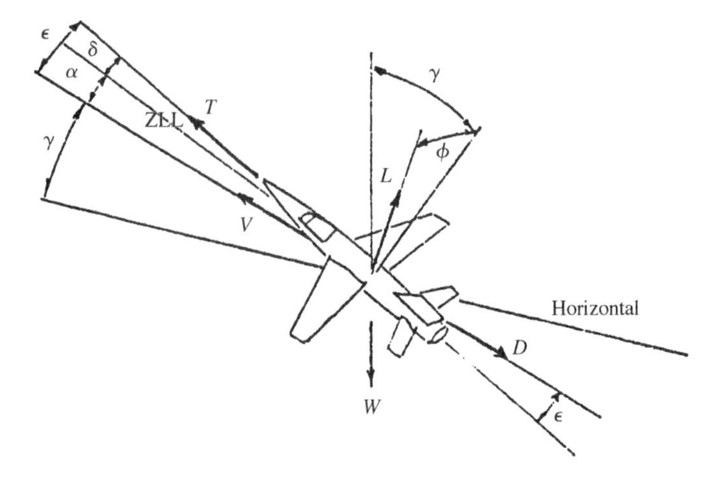

Figure 7.12 Climbing, Turning Flight

$$\tan \gamma = \frac{1}{L/D}$$

to give

$$\Delta h_\psi = \psi \frac{4(W/S)\cos^2 \gamma}{g\rho(C_L^2/C_D)\sin 2\phi} \tag{7.72}$$

To find the criteria for minimum height loss it can be assumed that the flight path is shallow; thus $\cos^2 \gamma = 1$. Also ψ can be replaced by 2π (one full turn). The height loss is clearly then minimized if the bank angle $\phi = 45°$, and $C_L = C_{L_{max}}$, which provides also the maximum condition for C_L^2/C_D. Eq. 7.72 shows also that an aircraft with a higher wing loading (W/S) loses weight more rapidly. As an aircraft descends, the density increases and the height loss is also decreased. A comparison can be made between the angle of glide γ in pure glide, where

$$\cot \gamma = \frac{L}{D} \tag{7.73}$$

and in turning glide where

$$\cot \gamma_\phi = \frac{L}{D}\cos \phi \tag{7.74}$$

It is easy to see that $\gamma_\phi > \gamma(\cot \gamma_\phi < \cot \gamma)$ as $\cos \approx 0.7$, and for pure glide $L/D = (L/D)_{max}$. For turning glide L/D corresponds to $C_{L_{max}}$, which yields a smaller L/D.

EXAMPLE 7.3

Consider a light aircraft with the following characteristics:

W = 3,000 lb
S = 225 ft^2
$C_{L_{max}} = 1.2$
$C_D = 0.025 + 0.06C_L^2$
$h = 5{,}000$ ft

The aircraft is in a 45° steady, banked, gliding turn

Calculate the glide performance for one full turn. In a pure glide mode, the glide angle would be

$$\gamma = \tan^{-1}\frac{1}{E_{\mathrm{m}}} = \tan^{-1} = \frac{1}{12.9} = 4.43°$$

In a banked turn, assume $C_L = C_{L_{\max}} = 1.2$, which gives an $L/D = 8.8$. The velocity in the glide can be calculated from Eq. 7.37 with

$$C_D = 0.025 + 0.06(1.2)^2 = 0.1364$$

$$\cot\gamma = 8.8 \times 0.707 = 6.22$$

$$\gamma = 9.13°$$

$$V = \sqrt{\frac{2W\sin\gamma}{\rho SC_D}}$$

$$= \sqrt{\frac{2 \times 3{,}000 \times \sin 9.13°}{0.002377 \times 0.8617 \times 225 \times 0.1364}}$$

$$= 123\ \frac{\text{ft}}{\text{sec}}$$

The radius of turn in a 45° bank is given by Eq. 7.70:

$$R = \frac{V^2\cos\gamma}{g\tan\phi} = 464\ \text{ft}$$

The loss of altitude in one complete turn is

$$S_\phi = 2\pi R\tan\gamma = 2 \times \pi \times 464 \times \tan 9.13° = 468\ \text{ft}$$

EXAMPLE 7.4

A high-performance aircraft has the following characteristics:

$$W = 30{,}000\ \text{lb}$$
$$T_{\max} = 20{,}000\ \text{lb}$$
$$S = 460\ \text{ft}^2$$

$$C_{L_\alpha} = 0.06/\text{deg}$$
$$C_{L_{max}} = 1.2$$
$$\delta = 3° \text{ (thrust angle)}$$
$$C_D = 0.0185 + 0.06C_L^2$$
$$M = 0.8$$

is in a steady 10° climbing turn under full thrust at 30,000 ft altitude. Calculate the load factor and the turn radius. The load factor n can be calculated from Eq. 7.68. Since ϵ is not known, the procedure must be carried out iteratively. Assuming first that $\epsilon = 0$, Eq. 7.64 gives

$$D = T - W \sin \gamma = 20{,}000 - 30{,}000 \sin 10° = 14{,}790 \text{ lb}$$

The drag coefficient follows with

$$C_D = \frac{D}{\gamma/2pM^2S} = \frac{D}{qS} = \frac{14{,}790}{0.7 \times 2{,}117 \times 0.2970 \times 0.64 \times 460} = 0.114$$

The lift coefficient can be found from

$$C_L = \sqrt{\frac{C_D - C_{D_0}}{k}} = \sqrt{\frac{0.114 - 0.0185}{0.06}} = 1.26$$

which requirement exceeds $C_{L_{max}}$. For successive approximation purposes, one can proceed and calculate the following:

$$\alpha = \frac{C_L}{a} = \frac{1.26}{0.06} = 21°$$

Thus

$$\epsilon = 21° + 3° = 24°$$

$$D = T \cos \epsilon - W \sin \gamma = 20{,}000 \cos 24° - 30{,}000 \sin 10° = 13{,}061 \text{ lb}$$

Then

$$C_D = 0.1$$
$$C_L = 1.17$$
$$\alpha = 19.4°$$
$$\epsilon = 22.4°$$

These values are sufficiently close. Calculating now n, from Eq. 7.68:

$$n = \frac{T \sin \epsilon + C_L qS}{W} = \frac{20{,}000 \sin 22.4° + 1.17 \times 1.3 \times 10^5}{30{,}000}$$

$$= 5.32$$

The bank angle comes from Eq. 7.68:

$$\phi = \cos^{-1} \frac{\cos \gamma}{n} = \cos^{-1} \frac{\cos 10}{5.32} = 79.3°$$

The radius can be calculated from (V = .8a = 800 ft/sec):

$$R = \sqrt{\frac{V^2 \cos^2 \gamma}{g(n^2 - \cos^2 \gamma)}} = \sqrt{\frac{800^2 \cos^2 10}{32.2(5.32^2 - \cos^2 10)}} = 3{,}687 \text{ ft}$$

PROBLEMS

7.1 An airplane has the following characteristics:

V = 120 m.p.h., EAS
h = 10,000 ft
W = 3,400 lb
W/S = 24 lb/ft^2
L/D = 10
T_a = 6,200 lb

It makes a 90° turn in 18 seconds maintaining altitude and incidence angle

Calculate the load factor, bank angle, radius of turn, and the thrust horsepower required.
Ans. ϕ = 31.3°; R = 2,520 ft; 160 HP.

7.2 An aircraft is flying at 20,000 ft at $M = 0.5$. The aircraft is banked into a 180° turn, which is supposed to take 29.24 seconds to complete.

Assume C_L = constant
$T_a = 6{,}200$ lb
$W = 26{,}000$ lb
$n_1 = 7.5$
$S = 375$ ft^2
$C_D = 0.015 + 0.1055C_L^2$
$C_{L_{max}} = 1.6$

Calculate the turn radius and show why this turn (can/cannot) be made.
Ans: R = 8,763 ft; cannot, since $D > T_a$.

7.3 Show that the turn radius in a coordinated climbing maneuver can be given by

$$R = \frac{V^2 \cos^2 \gamma}{g\sqrt{n^2 - \cos^2 \gamma}}$$

7.4 Show that the turn rate in a coordinated climbing maneuver can be given by

$$\dot{\psi} = \frac{\sqrt{n^2 - \cos^2 \gamma}}{V \cos \gamma}$$

7.5 Calculate the F-18A turn rates for the following conditions: $S = 400$ ft^2, $W = 38{,}000$ lb, $T/W = .486$, $C_D = .0245 + .13C_L^2$, $C_{Lmax} = 2.8$, $n_1 = 7.0$

a. The turn rate at its minimum turn radius

b. The maximum turn rate

c. The maximum instantaneous turn rate

d. The minimum velocity and the turn rate at which the instantaneous turn rate can be sustained, (i.e. the values at the edge of stall)
Ans: 8.7 deg/sec; 11 deg/sec; 28.6 deg/sec; 193 ft/sec; 7.94 deg/sec

7.6 If the aircraft in the Example 7.4 is climbing at a 10° angle but is maintaining a 60° bank angle, what is the resulting acceleration at full thrust?

7.7 Solve Example 7.4 if the aircraft is in a 10° descending flight.

7.8 The aircraft in Problem 7.2 loses all power and it will start gliding. In order to attempt to land at a nearby airport, the pilot makes a 180° turn. How much height is lost in that maneuver? Ans: 4498 ft.

7.9 The aircraft in Example 7.4 is cruising at $M = 0.8$ at 30,000 ft. The aircraft is then banked to 50° angle and full power is applied. How much height is gained if the aircraft keeps flying at the same speed through a 270° turn?

7.10 A 727-200 is traveling at an altitude of 4,000 ft with an approach speed of 300 ft/sec TAS. Heading is due north, wind is from the south at 20 mph.

a. What is the minimum separation for the flight path from (say, the edge of) the runway so that the aircraft can perform a 180° turn to be conveniently aligned with the runway for the final approach?

b. With the assumptions and data outlined below, how much altitude will the aircraft lose?

At the beginning of the turn, the pilot will maintain the thrust and attitude and put the aircraft into a 30° bank angle. It is reasonable to assume that the airspeed in the turn will remain constant. Wheels are up and the aircraft weighs 130,000 lb. For calculation purposes, use sea-level air properties. The rest of known data:
With partial flaps $C_D = .021 + .076C_L^2$, $C_{L_{\max}} = 2.2$; $C_{L_{\max,\text{land}}} = 2.6$, $S = 1{,}700$ ft^2. Three JT80-11 engines at 15,000 lb ea.
Ans: a. separation $= 2R = 9{,}840$ ft, b. $\Delta h = -270$ ft.

7.11 Two aircraft are engaged in air-to-air combat. Minimum (sustained) turn radius at sea level determines the winner.
Aircraft A. $C_D = .0225 + .11C_L^2$, $C_{L_{\max}} = 1.6$, $T/W = .5$, $S = 375$ ft^2, $W = 40{,}000$ lb
Aircraft B. $C_D = .0245 + .13C_L^2$, $C_{L_{\max}} = 2.8$, $T/W = .486$, $S = 400$ ft^2, $W = 38{,}000$ lb
Ans: A = 1,882 ft, B = 1,368 ft.

7.12 A jet transport aircraft at 300 mph is vectored into a holding pattern at a turn rate of 1.5 deg/sec. Calculate the turn radius and the required T/W. Aircraft data at 20,000 ft altitude:

$$W/S = 100 \text{ lb/ft}^2,\ S = 2{,}000 \text{ ft}^2$$

$$C_D = .015 + .06C_L^2,\ C_{L_{\max}} = 1.8$$

$$L/D|_{\max} = 16.67$$

Ans: $R = 16{,}800$ ft, $T/W = .074$

7.13 Construct the energy maneuverability diagram for F15C with the following characteristics at sea level:

$$W = 60{,}000 \text{ lb},\ S = 608 \text{ ft}^2$$

$$C_{L_{\max}} = 1.785,\ C_{D_0} = .023,\ k = .133$$

$$T = 50{,}000 \text{ lb},\ n_1 = 7.33$$

It can be determined from the diagram (or, also calculated more precisely) that: $V_c = 340$ KTAS, maximum instantaneous turn rate is 23 deg/sec at 345 KTAS, maximum instantaneous turn radius is 1450 ft at 345 KTAS, maximum sustained turn rate is 15 deg/sec at 250 KTAS, minimum sustained turn radius is 1,510 ft at 250 KTAS. If 6 g's are maintained at the corner speed ($P_s = -250$ ft/sec), then the altitude loss amounts to 260 ft/sec.

8

Additional Topics

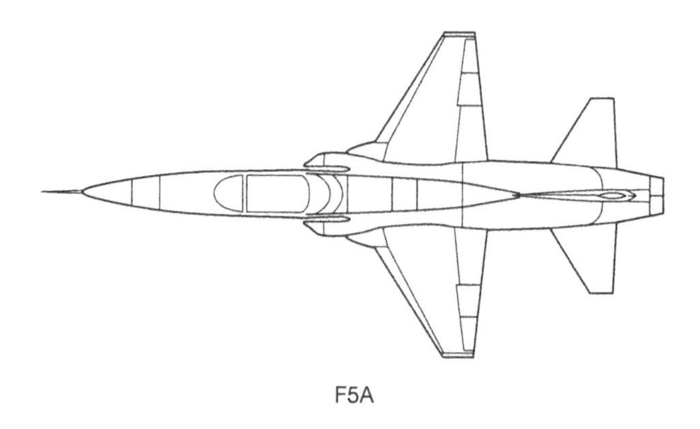

8.1 CONSTRAINT PLOT

One of the more useful and productive tools in mission analysis and preliminary design is the so-called constraint plot. Usually it is found in the format where take-off thrust loading T_o/W_o is plotted against the take-off wing loading W_o/S for mission requirements of interest. For a given set of mission specifications, called constraints, calculation of the range of typical T_o/W_o versus W_o/S values defines the available solution space for that particular set of missions (see Figure 8.1). A plot of T_o/W_o against W_o/S with all the applicable mission specifications as par-

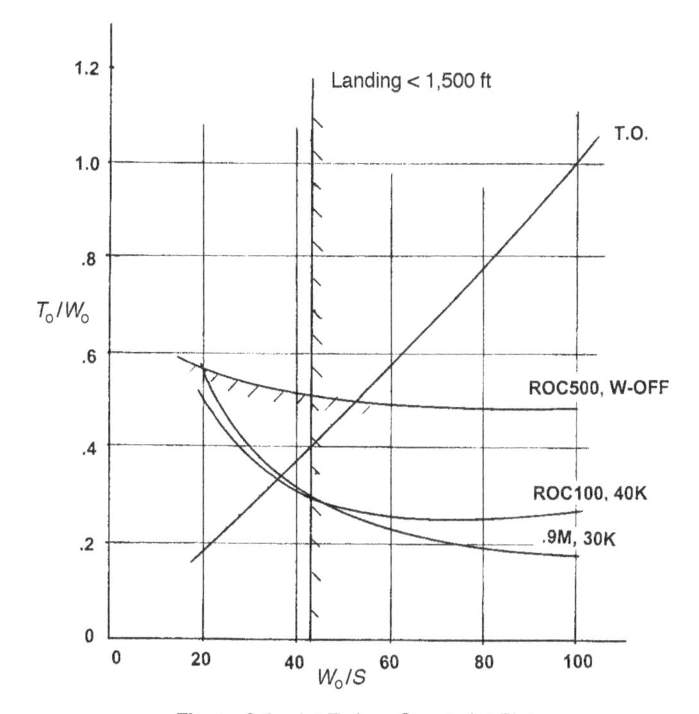

Figure 8.1 Jet Trainer Constraint Plot

ameters provides a map indicating a range of solutions available for the aircraft designer.

This provides one basis for choosing the *design point* of the aircraft. Since the design point seldom consists of a unique set of parameters, the constraint plot permits easy change or modification of constraints to establish trends and, via crossplotting unto carpet plots, render iteration accessible. Of course, the set of constraints (a wish list) may be established to be so overwhelming that, on one hand, they leave no room for choice and compromise of parameters—the essence of design. On the other hand, they may lead to physical conflicts (ultrashort runway) or exceeding existing technological limits (T/W exceeding, say, 5). In any case, a negative outcome here may yield also useful information.

Next, the most used mission constraints with their characteristic features will be listed in Table 8.1. Then the applicable equations used for evaluating thrust and wing loadings for each mission constraint will be reviewed. It is assumed that the general lift, drag, and thrust characteristics of the proposed aircraft are available. Finally, a numerical example will be used to demonstrate the constraint plot methodology.

TABLE 8.1 Typical Mission Constraints

Mission Specs	Characteristics	Parameters
Take-off	Ground run, rotation distance	V_T, T, T_r
Landing	Free roll, ground roll, braking	V_T, T, μ, ΔC_{D_0}
Air superiority	Energy rate P_s	M, T, altitude
Rate of climb	Rate of climb	M, T, altitude
Maximum speed	V or M specified	T, altitude
Acceleration	ΔM or g's specified	T, altitude
Sustained turning	(a) g-load (n)	M, altitude
	(b) turn rate	M, altitude

The equations used to calculate the thrust loading as a function of the wing loading (or the reverse) are the same ones as used to evaluate performance of the individual missions in the point performance sense. However, since some of the expressions and procedures (e.g., take-off run) may be rather lengthy and involved and in order to gain quick overview, usually simplified equations are used. This also includes many other simplifying assumptions concerning the variation of the main parameters with altitude and velocity. The emphasis in this methodology is on trends and relative behavior rather than on precision.

8.1.1 Take-off and Landing

Take-off *Take-off distance* is usually specified as

$$s_\mathrm{g} + s_\mathrm{r} \leq s_\mathrm{T}$$

which means that the ground run and the distance covered while the aircraft is in the rotation phase must be less that some prescribed total distance s_T. Ground run and rotation distances are given by

$$s_\mathrm{g} = \frac{1.44W/S}{\rho g C_{Lm} T/W}$$

$$s_\mathrm{r} = 1.2 t_\mathrm{r} V_\mathrm{s}$$

$$= 1.2 t_\mathrm{r} \sqrt{\frac{2W/S}{\rho C_{Lm}}}$$

Rotation time t_r must be specified. Combining the above equations yields

$$\frac{1.44.W/S}{\rho g C_{Lm} T/W} + 1.2t_r \sqrt{\frac{2W/S}{\rho C_{Lm}}} \leq s_T$$

This can be considered as a quadratic equation in $\sqrt{W/S}$ as a function of T/W, or, which is much simpler, solve the equation directly for the thrust loading T/W as

$$\frac{T}{W} = \frac{\dfrac{1.44W/S}{\rho g C_{Lm}}}{s_T - 1.2\sqrt{\dfrac{2W/S}{\rho C_{Lm}}}} \equiv \frac{aW/S}{s_T - b\sqrt{W/S}}$$

Landing *Landing* distance is usually specified as

$$s_{fr} + s_b \leq s_L$$

with the total distance s_L and free roll time t_r specified. Free roll distance s_{fr} may be expressed as

$$s_{fr} = 1.1t_r V_s.$$

Landing distance with braking and retarding devices may be expressed as

$$s_b = \frac{1.21W/S}{\rho g C_{Lm}\left[\dfrac{C_{D0} qS}{W} + \mu\right]}$$

During landing, V (in q) is evaluated at $.7V_T$, as $q = .5\rho(.7V_T)^2 = .245\rho(1.1V_s)^2$. C_{D0} and μ specifications must reflect expected braking configuration (e.g., brakes, chutes, spoilers, and runway conditions).

Normally, summation of free roll and braking distance yields a quadratic equation in $\sqrt{W/S}$ independent of T/W:

$$1.1t_r\sqrt{\frac{2W/S}{\rho C_{Lm}}} + \frac{1.21W/S}{\rho g C_{Lm}\left[\dfrac{C_{D0} qS}{W} + \mu\right]}$$

8.1.2 Constraints Tied to Performance Equation

The following six constraints will be evaluated by use of the fundamental performance equation (for the excess energy (rate)), which is then specialized for each individual requirement:

$$P_s \equiv \frac{de}{dt} = \frac{dh}{dt} + \frac{V}{g}\frac{dV}{dt} = \left(\frac{T - D}{W}\right)V$$
$$= Ma\left[\frac{T}{W} - \left(\frac{C_{D0}q}{W/S} + kn^2\frac{W/S}{q}\right)\right]$$

In the sequence of constraints 3 through 7, usually a full thrust is used. Thrust may be given by an equation of the form $T = T_o f(M)\sigma$, where $f(M)$ is normally curve-fitted to provide an expression like $(1 + aM)$, $(1 + aM^2)$, etc.

Air Superiority For air superiority, P_s, h, M are specified, $n = 1$, and the following equation results:

$$\frac{T}{W} = a + \frac{b}{W/S} + cW/S$$

Rate of Climb For rate of climb, set $P_s = dh/dt$ and $dV/dt = 0$, with h and M specified, $n = 1$. One obtains then again

$$\frac{T}{W} = a + \frac{b}{W/S} + cW/S$$

as in air superiority, only the constants a, b, c will take on different values.

Acceleration For acceleration, $P_s = dV/dt$, $dh/dt = 0$ and M and h are specified. One obtains the same type of equations as in the two previous constraints.

Maximum Speed For maximum speed conditions, set $P_s = 0$ (dV/dt and dh/dt are both zero), specify $V = V_{max}$ and h, $n = 1$. The following equation results:

$$\frac{T}{W} = \frac{b}{W/S} + cW/S$$

Again, the constants b and c will take on different values from previous constraints settings with:

$$b = C_{D0}q$$

$$c = \frac{k}{q}$$

Sustained Turning

1. Given g-load (n).
   Here, set $P_s = 0$ with n, M, and h specified. Again, the following equation is obtained:

   $$\frac{T}{W} = \frac{b}{W/S} + cW/S$$

   with the constants b and c:

   $$b = C_{D0}q$$

   $$c = \frac{kn^2}{q}$$

2. Turn rate given.
   Here, set $P_s = 0$ with M and h specified. Turn rate is given by $\dot{\psi}$ and the same equation as given g-load applies. n is now determined from

   $$\dot{\psi} = \frac{g}{V}\sqrt{n^2 - 1}$$

This completes a brief survey of the typical specifications used for establishing an aircraft-performance constraint plot. The list is not exhaustive. To explore the two primary design parameters, thrust-weight ratio and the wing loading, just about any aircraft-performance parameter may be turned into a requirement specification—turn time, range, endurance, and so on. Example 8.1 serves to illustrate intricacies and possibilities of this technique to focus on what is possible, what is needed, or what may be excessive.

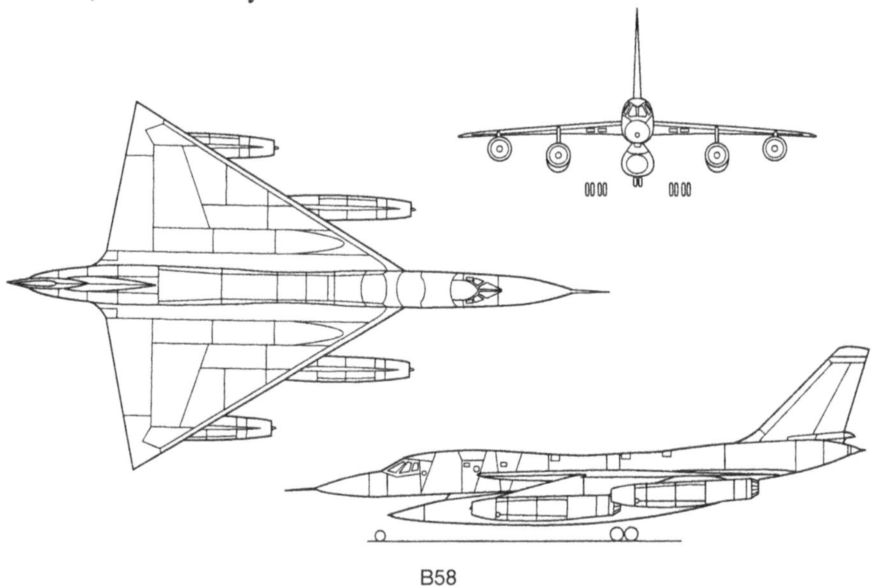

B58

EXAMPLE 8.1

In this example, the requirements for a new Navy jet trainer are explored. It is anticipated (from existing aircraft data) that the following aerodynamic properties should be applicable:

$$C_D = .018 + .15C_L^2, \quad C_{L\max} = 1.7$$

The following constraints are of interest:

1. Take-off distance at sea level not to exceed 1,500 ft at a temperature of 103°F. Include ground roll plus 1 second rotation time.
2. Maximum flight speed at 30,000 ft is $.9M$. Assume $W = .8W_o$, $T = T_o(1+M/2)\sigma$
3. Maintain a minimum rate-of-climb ROC = 100 fpm (service ceiling) at 40,000 ft, $.8M$, $W = .8W_o$, $T = T_o(1 + M/2)\sigma$.
4. Maintain a minimum ROC = 500 fpm under tropical day conditions (90°F) and wave-off condition, one engine out, at maximum landing $C_{D_0} = .1$, $W = .7W_o$, $T = T_o\sigma$.

5. Landing distance not to exceed 1,500 ft at 103°F. $W = .7W_o$, $C_{D_0} = .1$, brakes only $\mu = .3$, $L = T = 0$, 2 sec free roll.

The constraints will be considered in the same order as given in the requirements statement above. Note that only for simplicity, five are used from Table 8.1.

1. *Sea-level take-off* The only other item is to establish the air density at 103°F as

$$\rho = .002377(520/563) = .0022 \text{ sl/ft}^3$$

Thus, the landing equation will be used as given above:

$$\frac{T_o}{W_o} \le \frac{1.44\rho g C_{L\max} \dfrac{W}{S}}{s_T - 1.2t_r\sqrt{\dfrac{W/S}{\rho C_{L\max}}}}$$

$$= \frac{\dfrac{1.44}{.0022(32.2)1.7} W/S}{1500 - 1.2\sqrt{\dfrac{2}{.0022(1.7)}}\sqrt{W/S}}$$

$$= \frac{11.96W/S}{1500 - 27.75\sqrt{W/S}}$$

This permits setting up a table of values, $W = W_o$, $T = T_o$

2. *Level flight maximum speed, .9M, 30,000 ft* At 30,000 ft $\sigma = .375$, $V = Ma = .9(995) = 896$ ft/sec.

$$n = 1$$

$$q = \frac{1}{2}\rho V^2 = \frac{.0008907}{2} 896^2 = 357 \text{ lb/ft}^2$$

$$T = T_o(1.45).375 = .544T_o$$

Thus, the maximum speed constraint becomes

$$\frac{T}{W} = \frac{C_{D_0}q}{W/S} + \frac{k}{q} W/S$$

$$\frac{.544T_o}{.8W_o} = \frac{.018(375)}{.8W_o/S} - \frac{.15(.8)W_o/S}{357}$$

$$\frac{T_o}{W_o} = \frac{11.81}{W_o/S} + .000494(W_o/S),$$

which then yields

W_o/S	20	40	60	80	100
T_o/W_o	.6	.315	.226	.187	.168

3. *Maintain a minimum ROC = 100 fpm at 40,000ft* At 40,0000 ft $\sigma = .246$, $V = Ma = .8(968) = 774$ ft/sec.

$$n = 1$$

$$q = \frac{.002377}{2}(.246)774^2 = 176 \text{ lb/ft}^2$$

$$T = T_o(1.4).246 = .344T_o$$

$$P_s = \frac{dh}{dt} = \frac{100}{60} = 1.667 \text{ ft/sec}$$

The general constraint equation becomes

$$P_s = V\left[\frac{T}{W} - \frac{C_{D0}q}{W/S} - \frac{k}{q}W/S\right]$$

$$1.6677 = 774\left[\frac{.344T_o}{.8W_o} - \frac{.018(176)}{.8W_o/S} - \frac{.15}{176}.8W_o/S\right]$$

$$\frac{T_o}{W_o} = \frac{9.209}{W_o/S} + .00159W_o/S + .005$$

which yields the following set of values for the thrust and wing loading:

W_o/S	20	40	60	80	100
T_o/W_o	.5	.3	.254	.247	.256

4. *Maintain an ROC of 500 fpm at wave-off, one engine, sea level* At landing conditions $C_{D_0} = .1$, $W = .7W_o$, $\sigma = 520/550 = .945$. Speed of sound $a = 1{,}116/.945 = 1{,}148$ ft/sec.

$$n = 1$$

$$T \approx \frac{T_o}{2}\,\sigma = .4725T_o$$

$$V = 1.1V_s = 1.1\sqrt{\frac{2(.7)}{.002377(.945)1.7}}\,\sqrt{W/S} = 21.06\sqrt{W/S}$$

$$q = \frac{.002377}{2}\,.945(21.06)^2 W_o/S = .5W_o/S$$

$$P_s = \frac{dh}{dt} = \frac{500}{60} = 8.33 \text{ ft/sec}$$

Using again the general constraint equation:

$$P_s = V\left[\frac{T}{W} - \frac{C_{D0}q}{W/S} - \frac{k}{q}\,W/S\right]$$

$$8.33 = 21.06\sqrt{W/S} \times \left[\frac{.475T_o}{.7W_o} - \frac{.1(.5)W_o/S}{.7W_o/S} - \frac{.15(.7)W_o/S}{.5W_o/S}\right], \text{ or}$$

$$\frac{T_o}{W_o} = \frac{.582}{\sqrt{W_o/S}} + .4137$$

The corresponding set of thrust-wing loading values are

W_o/S	20	40	60	80	100
T_o/W_o	.544	.506	.489	.479	.472

5. *Landing distance,* $s_T \leq 1{,}500$ ft At landing conditions: $C_{D0} = .1$, $\mu = .3$, $L = T = 0$, $t_r = 2$ sec. Hot, 103°F day, $\sigma = \sigma_o(520/563) = .924\,\sigma_o$.

$$W = .7W_o$$

$$V_L = 1.1V_s = 1.1\sqrt{\frac{2(.7)W_o/S}{.002377(.924)1.7}} = 21.3\sqrt{W_o/S}$$

$$q = .245(.002377).924(21.3)^2 W_o/S = .244W_o/S$$

The landing constraint equation, from 2. above, becomes

$$1{,}500 \leq t_r V_L + \frac{1.21 W/S}{\rho g C_{Lm}\left[\dfrac{C_{D0}q}{W/S} + \mu\right]}$$

$$1{,}500 \leq 2(21.3)\sqrt{W_o/S} + \frac{1.21(.7)W_o/S}{.002377g1.7\left[\dfrac{.1(.244)W_o/S}{.7W_o/S} + .3\right]}$$

$$1{,}500 \leq 42.6\sqrt{W_o/S} + 19.44W_o/S$$

Solving the quadratic in $\sqrt{W_o/S}$ yields $W_o/S \leq 60.2$ a result that is independent of the thrust loading T/W. The five limiting curves resulting from the means part calculations have been plotted on Figure 8.1.

These curves, arising from the requirements set above, show that a minimum sea-level thrust–weight ratio of .5, at $W_o/S \approx 44$, is needed to satisfy the initial constraints. If now the wave-off constraint were eliminated, then the thrust–weight ratio would be reduced to about .34 and the wing loading would drop to about 35. Such a move would considerably alleviate the initial thrust requirements and, in turn, yield a lighter aircraft design.

Whether totally realistic or not, this example serves to demonstrate usefulness of the constraint plot in establishing practical specifications for new aircraft and then to evaluate the implications on design and technology.

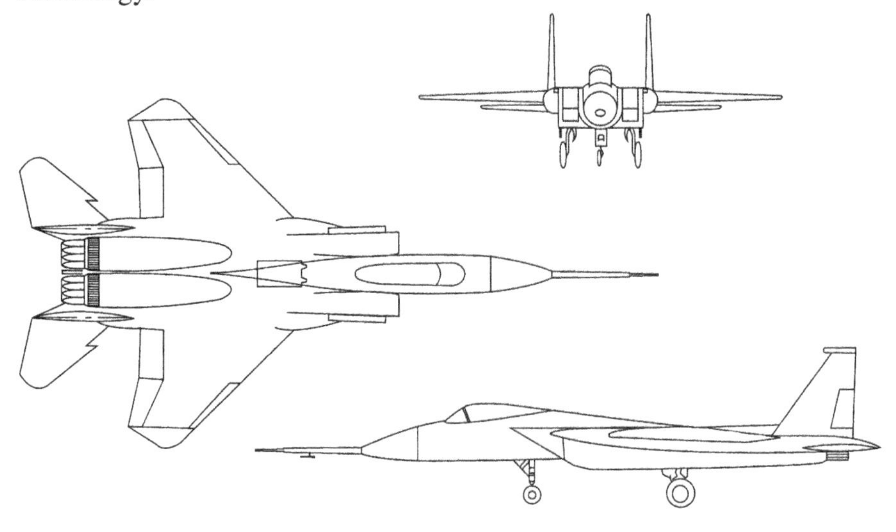

F15

8.2 ENERGY METHODS

The problem of determining aircraft performance optimum conditions (fastest climb, highest altitude, minimum fuel consumption, etc.) has been around since the advent of flight. For earlier low-performance aircraft, the standard point-performance techniques (discussed and used extensively in preceeding chapters) were sufficient to estimate the aircraft performance potential by using methods of single variable ordinary calculus. Faster and higher flying aircraft with more complex and supersonic missions have brought along the need to optimize the entire flight path. Thus, there are more variables to consider with attendant need for increased accuracy, and the point-performance (also called quasi-steady-state) methods must yield to an integral performance approach requiring use of calculus of variations or variational theory. This usually results in an increased level of calculational complexity and a loss of generality due to numerical computer solutions of nonlinear differential equations.

A simplified method, based on calculus of variations, is the energy-state (also called the energy-height) approximation, which has been shown to agree well with exact solutions. Here the aircraft is considered to be a point mass and its state is given in terms of its total energy, consisting of the sum of potential and kinetic energy. Now altitude and velocity are included in a single variable, energy, which then allows the use of the point-performance methods to build up approximate optimal trajectories.

The total energy concept was already introduced in Chapter 2 and used in Chapters 4 and 7 with the energy-maneuverability diagram. Aircraft specific energy is defined as

$$e \equiv \frac{E}{W} = h + \frac{V^2}{2g} = h_e, \quad \text{(energy height)} \tag{8.1}$$

and has the dimensions of length. Since h represents altitude, the units of h_e are given in feet or meters, and the specific energy is often called the energy height. It represents the theoretical altitude h that the aircraft could reach if all of the kinetic energy ($V^2/2g$) were converted (by a zoom) into potential energy. Figure 8.2 shows constant specific excess power and energy height curves on the altitude-airspeed map. Any aircraft at point A, at an altitude of 30,000 ft and at 400 knots (676 ft/sec) has an energy height of 37,088 ft because it is moving with a velocity of 676 ft/sec. Ideally, it could zoom to an altitude of 37,088 ft (with a final zero speed), or it could dive to sea level with a final

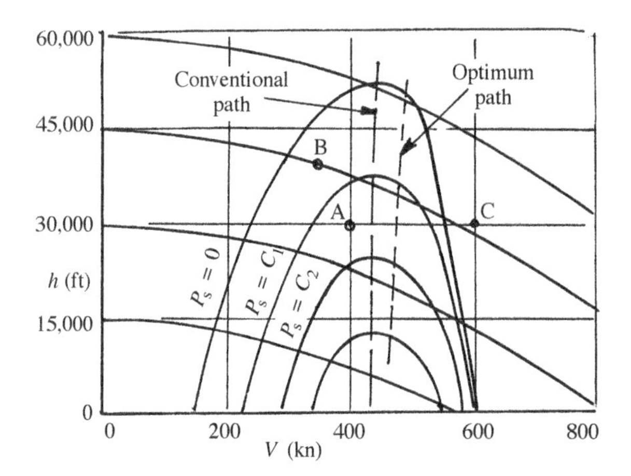

Figure 8.2 h_e and P_s Constant Lines; Conventional and Optimum Paths

(theoretical) velocity of 1,545 ft/sec (915 knots). In either case, its total specific energy remains at 37,088 ft.

Consider now points B and C on the h_e = 45,000 ft line. From point A to either points B or C there are many possible paths to follow but no matter which path is taken there will be a change of 7912 ft (45000-37088) in the energy height. Both B and C have the same specific energy state, both of which are at a higher level than point A. The ability of an aircraft to change its energy state is called energy maneuverability, as already seen in Chapter 7. The fundamental reason for the energy maneuverability method is to determine how to move from one energy state to another in some optimal manner—that is, the fastest time from A to B, minimum fuel flight from A to C, and so on. A minimum time transfer will be considered first, with other options introduced later.

The usefulness and need for the energy method will be demonstrated by considering flight from point A to point C, which requires an increase in the energy state. This can be accomplished by simply increasing the thrust to accelerate, at constant altitude, to a higher velocity. It will be seen later that it would be faster to dive first to a lower altitude where energy can be increased quickly and then zoom climb to reach point C.

The ability to change aircraft energy state comes from the fuel energy and is commonly expressed by the specific excess power

$$P_s \equiv \frac{P_a - DV}{W} = \frac{(T - D)V}{W} \tag{8.2}$$

which is equal to the rate of change of specific energy dh_e/dt. From Chapter 2, Eq. 2.20:

$$\frac{dh_e}{dt} \equiv \frac{dE/W}{dt} = \frac{dh}{dt} + \frac{V}{g}\frac{dV}{dt} = \frac{(T - D)V}{W} \equiv P_s \tag{8.3}$$

The left-hand side (kinematics) represents the sum of acceleration and the rate of climb and describes its maneuvering behavior. The right-hand side (aircraft specific power generation) defines the maneuvering capability (i.e., the energy needed to accelerate and/or climb, or move from, say, point A to point B or C). It should be noted that Eq. 8.3 is simply a restatement of the fundamental performance equation used in previous chapters.

Since P_s plays an important role in determining aircraft climb-acceleration capabilities, it is useful to briefly review the origin and meaning of the numerical values of the specific power. Specific excess power can be developed by use of Eq. 3.24 to yield

$$P_s = V\left[\frac{T}{W} - \frac{qC_{D0}}{W/S} - \frac{kn^2 W/S}{q}\right] \tag{8.4}$$

which shows that the specific power is also a function of the load factor. At a load factor of unity, if $P_s = 0$, thrust is equal to drag and a level flight is described. When plotted and superimposed for a typical aircraft on Figure 8.3, the result is a one-g level flight envelope. As the load factor increases, the values of P_s (specific excess power) will decrease with the attendant shrinking of the flight envelope. This is depicted in Figure 8.2 for the subsonic flight where the constants C1, C2, . . . represent either an increased g-level or a reduced value of P_s, or both. At the same time, a $P_s = 0$ contour defines, at a particular load factor, a constant-speed level altitude turning performance already studied

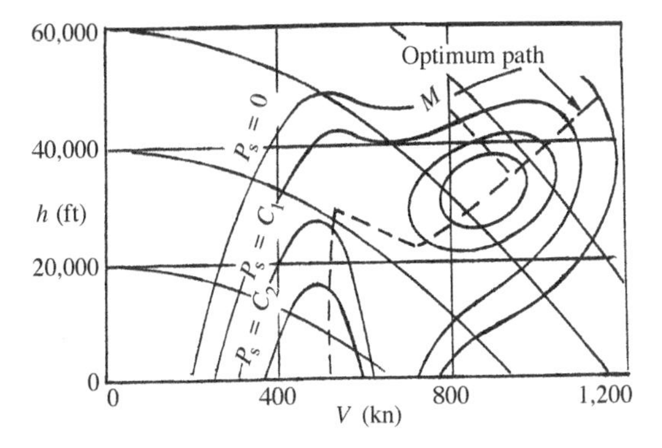

Figure 8.3 Minimum Time Path by Energy Method

with the doghouse plot in Chapter 7. When P_s is negative, the aircraft will lose altitude or decelerate. Local climb (zoom) may be accomplished, but at a loss of airspeed.

Consider now an F-18 aircraft flying at 30,000 ft with a velocity of 726 ft/sec, .7*M* (q = 334 lb/ft^2). Its weight is 38,000 lb, wing area is 400 ft^2, $C_D = .0245 + .13C_L^2$. This gives a lift coefficient $C_L = .285$ and a drag coefficient $C_D = .035$. If the local thrust/weight ratio is .4, the excess power is (from Eq. 8.2):

$$P_s = V\left(\frac{T}{W} - \frac{qC_D}{W/S}\right) = 726\left(.4 - \frac{334 \times .035}{95}\right) = 201 \text{ ft/sec} \quad (8.5)$$

The positive value of P_s means that the aircraft can accelerate or climb or a combination of both as long as the sum (the total specific energy rate) does not exceed 201 ft/sec. In this case, the instantaneous climb rate is

$$\left.\frac{dh}{dt}\right|_{V=c} = 201 \text{ ft/sec}$$

at one g and .7*M*. A level acceleration can be carried out at

$$\left.\frac{dV}{dt}\right|_{h=c} = \frac{P_s g}{V} = \frac{201 \times 32.2}{726} = 8.9 \text{ ft/sec}^2$$

A level flight cruise at .7M requires a reduction of thrust until $P_s = 0$. This clearly shows the usefulness of the specific energy in consolidating variables (h,V) and in representing aircraft performance capabilities.

Returning now to the problem of determining optimal paths, specifically to find the minimum time to change from one speed and altitude to another, the time can be expressed as (similar to Eq. 4.5)

$$t = \int_{h_{e1}}^{h_{e2}} \frac{1}{\frac{dh_e}{dt}}\, dh_e = \int_{h_{e1}}^{h_{e2}} \frac{1}{P_s}\, dh_e \quad (8.6)$$

where

$$\frac{dh_e}{dt} = \frac{(T - D)V}{W} = f(V, h) = P_s$$

The problem now becomes one of minimizing the above integral. The calculus of variations provides a simple condition for minimizing this integral (Rutowski):

$$\frac{\partial}{\partial V}\left(\frac{dh_e}{dt}\right)_{h_e=c} = 0 \quad \text{or} \quad \frac{\partial}{\partial h}\left(\frac{dh_e}{dt}\right)_{h_e=c} = 0$$

or

$$\frac{\partial}{\partial V}\left[\frac{(T-D)V}{W}\right]_{h_e=c} = 0 \quad \text{or} \quad \frac{\partial}{\partial h}\left[\frac{(T-D)V}{W}\right]_{h_e=c} = 0 \quad (8.7)$$

Equation 8.7, in any of the equivalent forms, is a requirement that the minimum time path satisfy this condition at every point. An analytical expression for the path cannot be defined in general, but the condition itself is sufficient to provide a solid basis for determining a minimum time path using the energy state concept.

This is easiest seen if one returns to the conventional method (Chapter 4) where the minimum time (maximum rate of climb) condition satisfies the following condition:

$$\frac{\partial}{\partial V}\left[\frac{(T-D)V)}{W}\right]_{h=c} = 0 \quad (8.8)$$

Equation 8.8, simply stated, implies that, at a given altitude, maximum specific excess power (P_s) defines the best climb rate. See also Figure 8.2, where the meaning of Eq. 8.8 is that the lines of *constant altitude* are tangent to P_s = const curves (see also Figure 4.5). The implications from conditions given by Eq. 8.7 are similar. The locations of the tangent point of *constant* h_e *curves* to the P_s = const curves define the minimum time path—that is, the optimum velocity at a given altitude, which also amounts to operating at maximum excess energy per pound of aircraft.

At lower altitudes, the path given by the energy state method shown in Figure 8.2 differs little from the conventional one but starts deviating at higher speeds and altitudes. The difference is seen by comparing the tangency conditions of the conventional and optimal paths. At supersonic speeds, the differences become more pronounced. Indeed, the conventional process, as given by Eq. 8.8 (see also Figure 4.5), is not appropriate because of the large speed changes involved, as indicated in Figure 8.3. The energy method is uniquely suited for this case.

Following Figure 8.3, a supersonic aircraft initially climbs at subsonic speed until it reaches the intersection of an h_e = constant line, which is tangent to two equal-valued P_s lines (the dashed line goes through two C_2's). A quick dive follows at constant energy (h_e) until the high-speed portion of C_2 is reached. Then a climb at increasing Mach number is accomplished according to Eq. 8.7 following the $h_e - P_s$ tangency points to a desired altitude and speed, or as defined by the dynamic pressure-operating envelope limit. Since the aircraft is then within the operating envelope ($P_s > 0$), higher altitudes to, say, point M at $P_s = 0$ can be reached by a (constant h_e) zoom; losing some speed and gaining altitude.

A shortcoming of the energy optimization method, for the aircraft shown in Fig. 8.3, is that the conversion of energy (from potential energy to kinetic in dive, reverse in zoom) is theoretically accomplished in zero time due to $\Delta h_e = 0$ in Eq. 8.9 along constant energy lines. The near-sonic dive portion to the P_s bubble appears mainly in aircraft with marginal sonic and supersonic excess thrust, and some judgment must be exercised in establishing the dive path. Then time can be estimated by elementary kinematic techniques.

For modern high-performance aircraft with more powerful and lighter engines, the subsonic-supersonic conversion via a dive may not appear, since the excess energy (P_s) is given by a rather regular layer of curves ascending to higher altitudes and toward $P_s = 0$, Figure 8.4 (see Example 8.2). The optimum path then simply follows the $h_e - P_s$ tangency points (the dashed lines) according to Eq. 8.7. However, a zoom technique may still be necessary to reach higher altitudes above the curve defining the optimum flight path.

The overall path time from one energy level (h_1, V_1) to another (h_2, V_2) can be evaluated by one of the following two methods. The first one is a simple approximation of Eq. 8.6 by a summation process

$$t = \int_{h_{e1}}^{h_{e2}} \frac{1}{\frac{dh_e}{dt}} dh_e = \sum \frac{1}{dh_e/dt} \Delta h_e = \sum \frac{\Delta h_e}{P_{savg}} \tag{8.9}$$

which is a process analogous to the one used with the standard climb technique given by Eq. 4.5. Now the integration limits are set by h_{e_2} and h_{e_1} rather than altitudes h_2 and h_1.

The second approach arises from the integral side of Eq. 8.9. Figure 8.5 shows the integrand ($1/P_s$) plotted as a series of curves for constant altitude between the limits h_{e_1} and h_{e_2}. Minimizing the integrand is

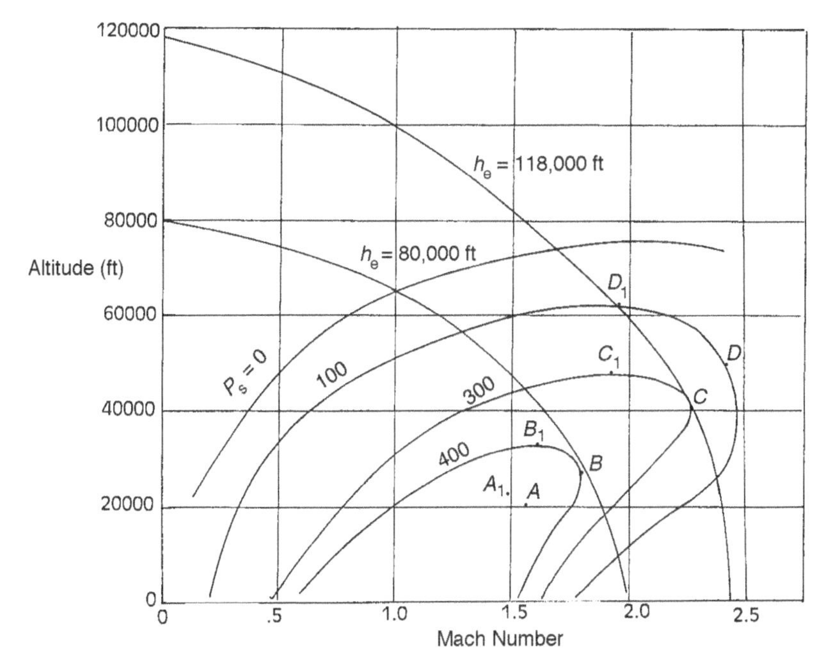

Figure 8.4 Excess Energy Curves

accomplished by selecting the points on the constant altitude curves according to Eq. 8.7 (i.e., at tangency points to the constant altitude curves). A sequence of such points establishes the optimum path envelope (shown by the short dashed lines), and the area under the envelope represents the minimum value of the time integral.

The standard climb curve, according to Eq. 8.8 (longer dashed lines), is also shown as the locus of the minimum points at constant altitude curves. The area under this curve is somewhat larger that the one corresponding to the optimum climb. Thus, the aircraft flying the optimum path reaches the same energy height before the aircraft using the standard climb technique. As previously pointed out, the difference in areas (climb time) is small at lower altitudes and speeds, but it shows a substantial increase at higher altitudes. The running time and the difference in the time for the energy path and the conventional path is shown in the upper part of Figure 8.5.

For commercial and general aviation aircraft, flying at more modest speeds and altitudes, the energy advantage is of no importance. The question of how to get from point A to point B, or to climb from takeoff to cruise altitude, with a minimum fuel consumption, is an item of much more consequence.

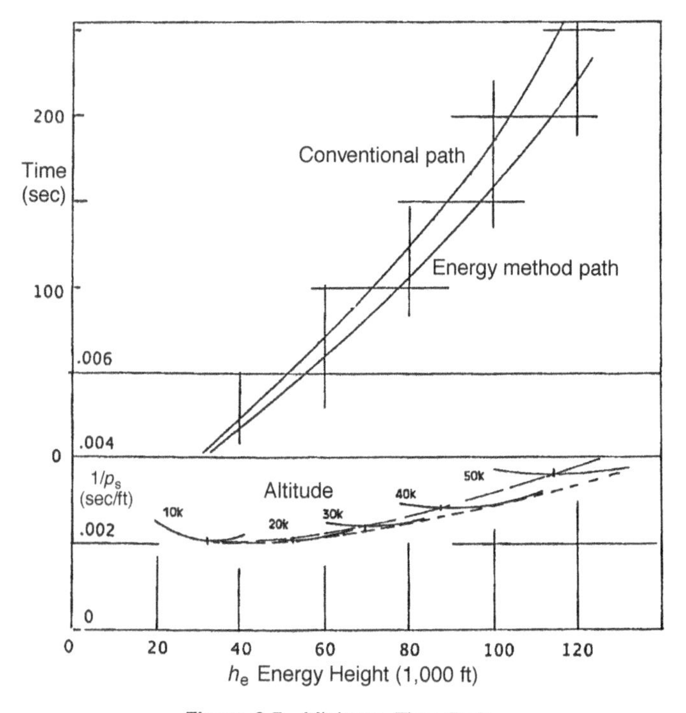

Figure 8.5 Minimum Time Paths

The amount of fuel required to fly from one energy level h_{e_1} to h_{e_2} is given by

$$w_f = \int_{h_{e1}}^{h_{e2}} \frac{dw_f}{dh_e} dh_e = \int_{h_{e1}}^{h_{e2}} \frac{dw_f}{dt} \frac{1}{\frac{dh_e}{dt}} dh_e \tag{8.10}$$

Since the rate of fuel consumption dw_f/dt is usually expressed as

$$\frac{dw_f}{dt} = CT$$

where C is the thrust specific fuel consumption and is usually a function of speed and altitude, then Eq. 8.10 can be written as

$$w_f = \int_{h_{e1}}^{h_{e2}} \frac{CT}{P_s} dh_e \tag{8.11}$$

This integral is minimized (the condition of minimum fuel path) when

$$\frac{\partial}{\partial h}\left(\frac{P_s}{CT}\right)_{h_e=c} = 0 \quad \text{or} \quad \frac{\partial}{\partial V}\left(\frac{P_s}{CT}\right)_{h_e=c} = 0 \tag{8.12}$$

which is similar to Eq. 8.7.

A typical minimum fuel path solution is shown in Figure 8.6 with some of the computational details given in Example 8.2. The same aircraft is used for both minimum time and minimum fuel problems. The amount of fuel can be calculated by the same technique as used in Figure 8.5 but one plots now $CT/P_C \equiv 1/f_s$ against h_e for constant altitude as a parameter. The area under the envelope curve represents the amount of fuel used.

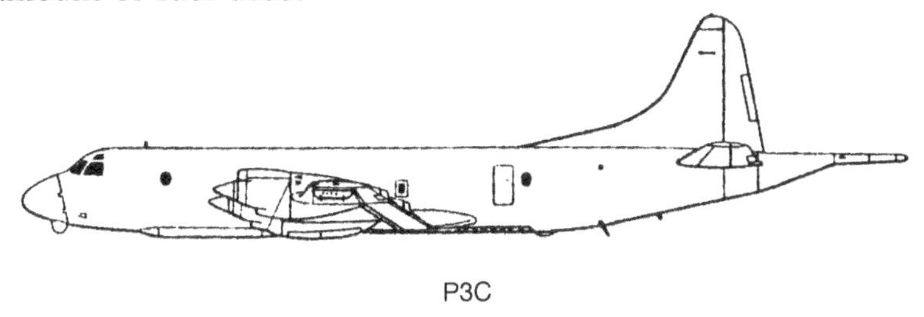

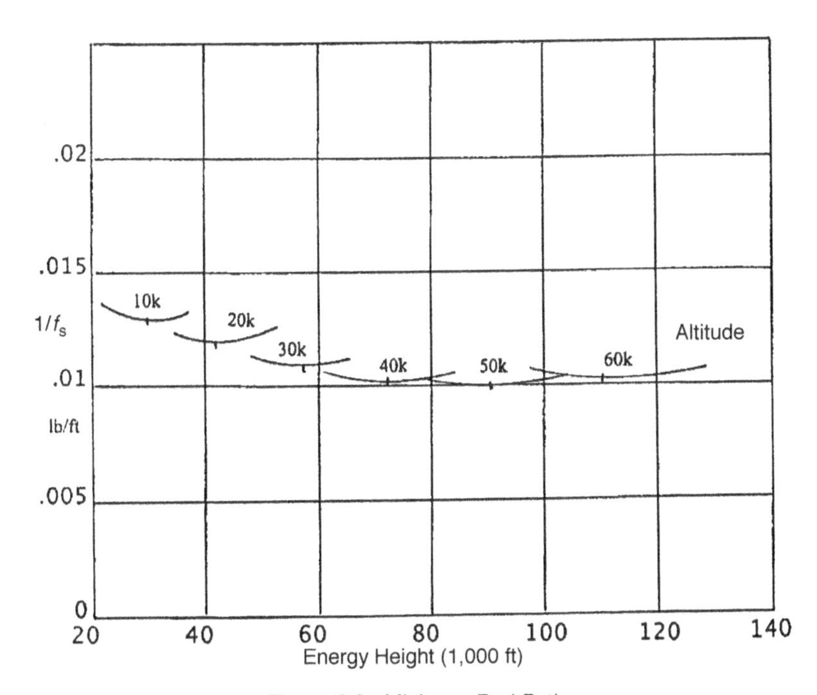

Figure 8.6 Minimum Fuel Path

EXAMPLE 8.2

Consider a hypothetical supersonic capable aircraft with the following characteristics:

$$C_D = C_{D_0} + kC_L^2,\ C_{D_0} = .016$$

$$k = 1(\pi \mathcal{R} e),\ e = .8M/6$$

$$S = 480\ \text{ft}^2,\ \mathcal{R} = 5$$

$$T = T_o(1 + .5M)\sigma^m,\ T_o = 20{,}000\ \text{lb}$$

$$m = .7\ h < 40{,}000\ \text{ft}$$

$$m = .8\ h \geq 40{,}000\ \text{ft}$$

$$C = C_o\sqrt{\theta},\ C_o = .85(1 + M/6)\ \text{lb}_m/\text{lb}_f/\text{hr}$$

The purpose of this excercise is to determine:

a. Minimum time flight path
b. The time along this flight path
c. Minimum fuel flight path

a. Minimum time path is obtained in three steps.

1. Calculate, for a number of altitudes, P_s as a function of Mach number from Eq. 8.3:

$$P_s = \frac{(T - D)V}{W}$$

   and plot P_s against Mach number for a number of altitudes.
2. Obtain constant P_s curves by crossplotting the P_s values, at constant altitude, against Mach number, as shown in Figure 8.4.
3. Obtain the minimum time energy path from the h_e = constant tangency points to the P_s = constant curves.

Several such points are shown. One (B), where h_e = 80,000 and the other (*C*), for h_e = 118,000 constant curves are tangents to P_s = 400 and 300 curves, respectively. The others are found in Tables 8.1 and 8.2.

TABLE 8.2 Standard Flight Path, Time

Point	P_s	h_e	Alt.	M	Δt
A_1	469	54900	23k	1.45	
B_1	410	69300	30k	1.6	
C_1	300	98000	43.5k	1.94	
D_1	100	117000	61k	1.94	

b. The time along the flight path, from one altitude to another, is obtained by plotting $1/P_s$ curves, at constant altitudes against h_e. The numerical values have been computed already in a. The calculations can be carried out by use of Eq. 8.9 in the difference form or the integral form. Here, both have been used. The results are shown in Figure 8.5.

The area under the short dashed lines gives the time for the optimum path between any desired energy height h_e. The longer dashed lines give the time computed under the conventional method. The short dashed lines represent the envelope drawn to the constant altitude curves which represents time for the optimum climb path.

To assist in obtaining and verifying the results for the optimum and standard fight paths, the pertinent data is given in the tables below. In both tables columns are shown for P_s, h_e, altitude, and Mach number at the four points, also shown in Figure 8.4. The time column is left open to be filled out by the reader.

Table 8.2 contains the data for the path to be calculated by the standard techniques (Chapter 4). The calculation procedure is given for both the standard method and the optimal path by Eq. 8.9. For the standard method one uses the value of h_e as found by the constant altitude tangent line to a P_s constant line (i.e., at the points A_1 . . . D_1. The optimal path is calculated from the data found at points A . . . D.

Table 8.3 lists the data for the optimal path points.

TABLE 8.3 Optimal Path Points, Time

Point	P_s	h_e	Alt.	M	Δt
A	450	62800	20k	1.6	
B	400	80000	30k	1.8	
C	300	118000	40k	2.3	
D	100	131000	43.5k	2.44	

TABLE 8.4 Optimal Path Points, Fuel

Alt.	$1/f_s$	h_e	Δlb
10k	.01295	30,000	
20k	.01295	42,000	
30k	.01095	57,500	
40k	.0106	72,000	
50k	.01024	90,500	

The results obtainable from both tables, as compared to Figures 8.4 and 8.5, may not agree precisely due to roundoffs in the tables and due to plotting of lines with high curvature to obtain tangent points.

When calculating path time from Tables 8.1 and 8.2 (or any other data source), extreme care should be taken in interpreting the results. The energy method delivers the results, shows advantage, between specific energy values—but not between two altitudes. For example, the times between points A_1 and C_1 (standard) and between A and C (optimal) turn out to be (about) 115 and 146 sec, respectively. Even if the optimal path aircraft spends only, say 4 seconds to climb to same 43,500 ft altitude, it is still 30 seconds behind. However, it has a much higher specific energy due to higher Mach number. This may be of tactical advantage.

When comparing same specific energy end points, D_1 and C, then the standard-path aircraft needs another 100 sec to get to point D_1 and the tables are turned in favor of the optimal path approach.

c. Minimum fuel path can be calculated by plotting $1/f_s = CT/P_s$ against the energy height h_e, as shown in Figure 8.6. Area under the envelope drawn to the constant altitude curves gives the fuel used between any energy heights. It turns out that the envelope curve follows very closely the minimum points on the curves. Since it is much easier to locate the minimum points during calculations, it is practical to carry out the minimum fuel calculations on the curve drawn through the minimum points, rather than through the tangent points.

c. To facilitate following through with some numerical results, Table 8.4 gives the values for $1/f_s$ for a number of altitudes shown in Figure 8.6. θ is the local atmospheric temperature ratio.

PROBLEMS

8.1 In Example 8.2, show the energy path from sea level at $M = 1.0$ to $P_s = 0$ at $M = 2.0$.

8.2 In Example 8.2, calculate and compare the flight times for the energy path and the one obtained by conventional method between 20,000 ft and 40,000 ft. Compare your results with the value obtainable from Figure 8.5.

8.3 In Example 8.2, calculate the amount of fuel used by a climb from 20,000 ft to 50,000 ft.
Ans: 660 lb.

A

Properties of Standard Atmosphere

U.S. Standard Atmosphere, 1962

h ft	T °R	p lb/ft^2	δ	σ	a ft/sec	$\mu \times E7$ slug/ft/sec	$\nu \times E4$ ft^2/sec
−2000	526	2274	1.074	1.060	1124	3.75	1.49
−1000	522	2193	1.037	1.030	1120	3.73	1.52
0	519	2116	1.000	1.000	1116	3.72	1.56
1000	515	2041	.964	.971	1113	3.70	1.60
2000	512	1968	.930	.943	1109	3.70	1.64
3000	508	1897	.896	.915	1105	3.66	1.68
4000	504	1828	.864	.888	1101	3.64	1.72
5000	501	1761	.832	.862	1097	3.62	1.77
6000	497	1696	.801	.836	1093	3.60	1.81
7000	494	1633	.772	.811	1089	3.58	1.86
8000	490	1572	.743	.786	1085	3.56	1.90
9000	487	1513	.715	.762	1081	3.54	1.95
10000	483	1455	.688	.739	1077	3.52	2.00
11000	480	1400	.661	.716	1073	3.50	2.05
12000	476	1346	.636	.693	1069	3.47	2.11
13000	471	1294	.611	.671	1065	3.45	2.16
14000	469	1243	.588	.650	1061	3.43	2.22
15000	465	1194	.564	.629	1057	3.41	2.28
16000	462	1147	.542	.609	1053	3.39	2.34
17000	458	1101	.520	.589	1049	3.37	2.40
18000	454	1057	.499	.570	1045	3.35	2.54
19000	451	1014	.479	.511	1041	3.33	2.61
20000	447	973	.460	.533	1037	3.31	2.61
21000	444	933	.441	.515	1033	3.28	2.68

h ft	T °R	p lb/ft^2	δ	σ	a ft/sec	$\mu \times E7$ slug/ft/sec	$\nu \times E4$ ft^2/sec
22000	440	894	.422	.500	1029	3.26	2.76
23000	437	856	.405	.481	1024	3.24	2.83
24000	433	820	.388	.464	1020	3.22	2.92
25000	430	785	.371	.448	1016	3.20	3.00
26000	426	752	.355	.433	1012	3.17	3.09
27000	422	719	.340	.417	1008	3.15	3.18
28000	419	688	.325	.417	1003	3.13	3.27
29000	415	658	.311	.388	989	3.11	3.37
30000	412	629	.297	.374	995	3.09	3.45
31000	408	600	.284	.361	990	3.06	3.57
32000	405	573	.271	.347	987	3.04	3.68
33000	401	547	.259	.335	982	3.02	3.80
34000	397	522	.247	.322	977	3.00	3.93
35000	394	498	.235	.310	973	2.97	4.04
36000	390	475	.224	.298	969	2.95	4.18
37000	390	453	.214	.284	968	2.96	4.38
38000	390	432	.204	.271	968	2.96	4.59
39000	390	411	.194	.258	968	2.96	4.82
40000	390	392	.185	.246	968	2.96	5.06
41000	390	373	.176	.235	968	2.96	5.31
42000	390	356	.168	.224	968	2.96	5.57
43000	390	339	.160	.213	968	2.96	5.84
44000	390	323	.153	.203	968	2.96	6.13
45000	390	308	.146	.194	968	2.96	6.43
50000	390	242	.115	.152	968	2.96	8.18
55000	390	190	.090	.120	968	2.96	10.41
60000	390	151	.071	.094	968	2.96	13.22
65000	390	118	.056	.074	968	2.96	16.83
70000	392	93	.0430	.058	971	2.96	21.551
75000	395	73	.0345	.045	974	2.96	27.67
80000	398	58	.027	.036	978	2.96	35.05
85000	401	46	.022	.028	981	2.96	44.63
90000	403	36	.017	.022	985	2.96	56.71
95000	406	29	.014	.017	988	2.96	71.98
100000	409	23	.011	.014	991	2.96	91.22

h altitude
p pressure
μ coefficient of viscosity

T temperature
$\delta = p/p_o$ pressure ratio
$\nu = \mu/\rho$ coefficient of kinematic viscosity
$\theta = T/T_0$ temperature ratio

a speed of sound
$\sigma = \rho/\rho_o$ density ratio
$p_o = 14.7$ *psia*
$\rho_o = .002377$ *slugs/ft*^{3}
$T_0 = 519$°R

B

On the Drag Coefficient

Aircraft drag can be expressed as

$$D = \frac{1}{2}\rho V^2 C_D S$$

where C_D is the aircraft total drag coefficient. A typical set of curves of the drag coefficient is shown in Figure B.1 as a function of the Mach number for selected lift coefficients. It is seen that the drag coefficient remains nearly constant in the subsonic region up to the critical Mach number and starts increasing rapidly in the transonic range after reaching the drag divergence Mach number. In the supersonic range the drag coefficient tends to decrease rapidly.

The level of C_{D0} depends on the type of aircraft (glider, fighter, transport, etc.). For some typical data of C_{D0} see Appendix C.

The drag coefficient C_D plays a crucial role in aircraft performance evaluation because it represents most of the aircraft aerodynamics in its variation with the flight parameters.

In its accurate and complete representation, the drag coefficient may be given in a graphical (see Figure B.1) or piecewise polynomial form to accommodate both the required lift and Mach number ranges. For practical engineering calculations where analytical results are desirable, the following expression is used for the drag coefficient of an airplane:

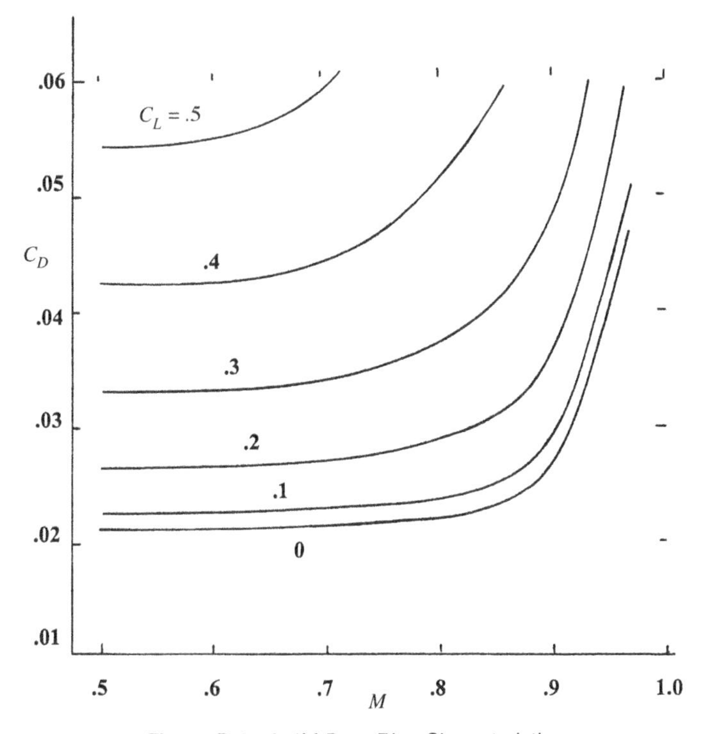

Figure B.1 A-4M Drag Rise Characteristics

$$
\begin{aligned}
C_D &= C_{D_0} + C_{D_L} \\
&= C_{D_0} + kC_L^2
\end{aligned}
\tag{B.1}
$$

where C_{D_L} is the drag due to lift and C_{D_0} is the zero lift drag coefficient due to parasite (viscous + form) drag. Due to its parabolic shape and due to its early representation in polar form, Eq. B.1 has acquired the label drag polar.

Eq. B.1 is a direct consequence of finite wing theory results and works best for uncambered aircraft in low subsonic range. The coefficient k is given by

$$
k = \frac{1}{\pi ARe}
$$

The Oswald span efficiency factor e is a function of the aspect ratio, increasing with increasing aspect ratio. Typical values range from .6 to .9, but it becomes a function of the Mach number and C_L near and

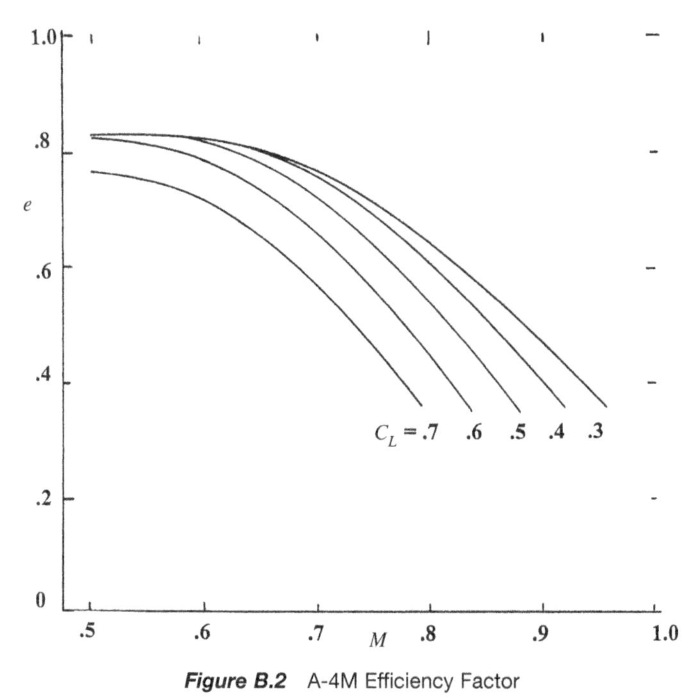

Figure B.2 A-4M Efficiency Factor

above the critical Mach number where it exhibits a continuing decrease to about 50 percent of its low speed value, Figure B.2.

Although the efficiency factor e depends on a number of parameters, the following correlation for wing-body combinations should serve as a first order of approximation:

$$e = 4.61(1 - .045AR \cdot 68)\cos \Lambda \cdot 15 - 3.1$$

where Λ should be determined by the locus of $t/c|_{\max}$ for the wing. In practice, leading-edge value Λ_{LE} is used for quick estimation.

Typical drag polar at subsonic speeds for A-7E aircraft is shown in Figure B.3. At supersonic speeds the form of Eq. B.1 is still valid but the wave drag effects need to be included in both k and C_{D0} due to their lift and nonlift contributions, respectively.

Supersonic drag can be estimated from

$$\text{Total drag} = \text{Parasite drag} + \text{Wave drag} + \text{Induced drag}$$

or

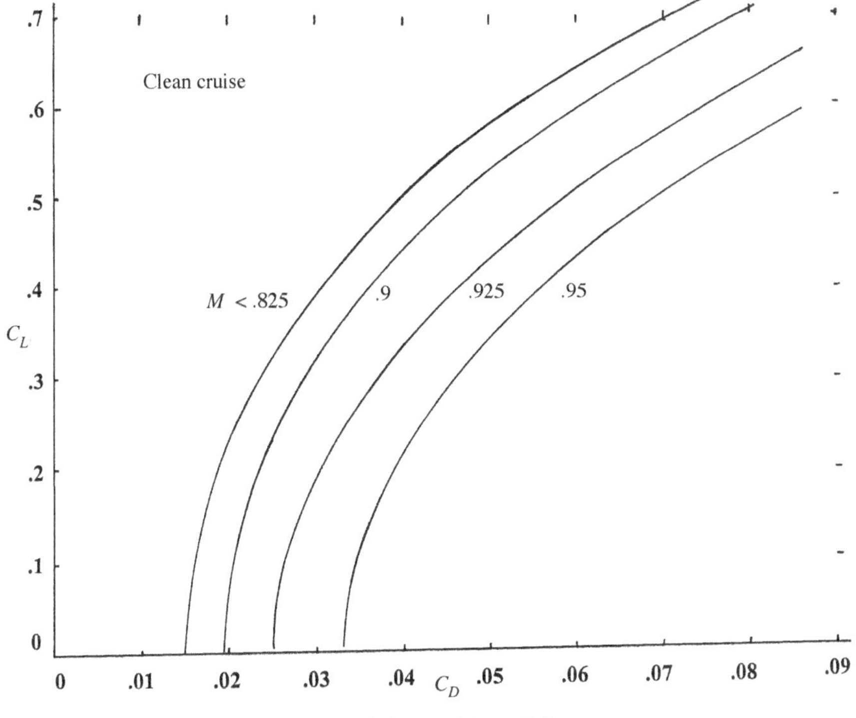

Figure B.3 A-7E Drag Polar

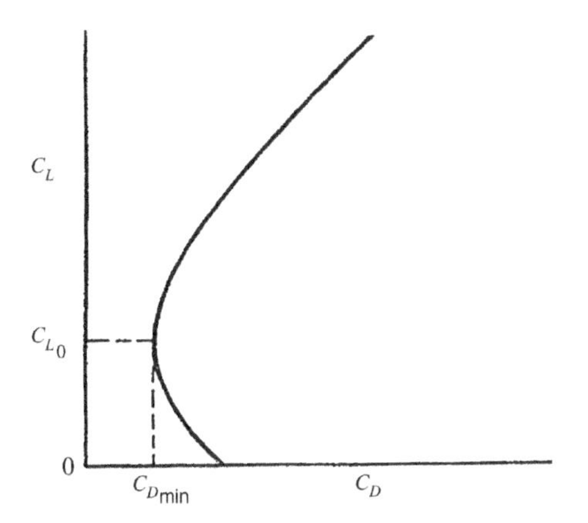

Figure B.4 Cambered Aircraft Polar

$$C_D = C_{D_p} + \text{Zero lift wave drag} + \text{Wave drag due to lift}$$
$$+ \text{Induced drag}$$

At its simplest, it can be written as

$$C_D = C_{D_p} + \frac{49(t/c)^2}{\sqrt{M^2 - 1}} + \frac{\sqrt{M^2 - 1}}{4} C_L^2 + kC_L^2$$
$$= C_{D_0} + C_{D_{0w}} = (k_1 + k)C_L^2$$
$$= C_{D_0} + KC_L^2$$

An improved version of k_1 in supersonic flight is

$$k_1 = \frac{AR(M^2 - 1)}{4AR\sqrt{M^2 - 1} - 2} \cos \Lambda$$

In transonic region, $K \equiv k$ transitions slowly from k to k + k_1. In transonic region, beyond the critical Mach number, more advanced correlations must be used to estimate the zero lift drag rise to low supersonic Mach numbers.

For an aircraft with camber (or with flaps extended), the drag polar will still be a parabola, but it will be shifted up along the lift axis, as shown in Figure B.4. Thus, the drag polar is given by a simple translation:

$$C_D = C_{D_{\min}} + k(C_L - C_{L_0})^2 + k_1 C_L^2 \qquad \text{(B.2)}$$

For many aircraft with relatively low camber, the difference between C_{D_0} and $C_{D_{\min}}$ is small and may be neglected. Thus, C_{D_0} is usually used in practice. For clarity and simplicity, Eq. B.1 is used throughout most of the development in this book.

C

Selected Aircraft Data

Aircraft	W_m	W_e	S	AR	C_{D_0}	k	e	L/D_m	R	C_{L_m}	W_f
767-200	320,000	177,000	3,080	7.9	.018		.8		5,200		112,000
747-200	830,000	380,000	5,500	7	.0148		.85	17.7	6,800	2.5	310,000
737-200	139,000	73,700	1,140	7.9	.019			17.6	2,700	3.2	34,500
DC10-30	555,000	241,000	4,600	7.6	.0162			17.7	4,700	2.5	
A330-200	507,000	266,000	3,900	10	.018			17.5	5,600		247,000
DC3	25,000	17,700	987	9.14	.0249		.75	14.7			
DHC8	34,500	23,000	585	12.4	.02	.0322	.8			1.5	5,700
A-4M	24,500	10,500	260	2.91	.0213	.132	.82		1,900		5,440
A-7	42,000	19,000	375	4	.015	.1055	.87		2,250	1.6	10,000
F4E	62,000	29,000	530	2.8	.0224			8.6	1,900		
F5A	21,000	8,100	170	3.8						2	
F15C	68,000	28,000	608	3	.023	.133				1.785	
F16A	35,400	14,600	300	3	.018					1.65	
F18A	51,000	24,000	400	3.5	.0243	.13				2.8	
F14A	74,000	41,000	565	7.3/1.9	.0255						
Rafale	40,000		375		.0225	.11				1.6	
MirageIII	30,000	16,000	375	1.9	.013					1	
Piper Cherokee	2,450	1,400	170	6	.0358		.76	10		1.7	

Notes. Most of the data in this table should be considered as approximate, or average, and may cross the specifications of several configurations. For example, the maximum weight W_m is a function of the aircraft loading and the amount of fuel on board. Thus, it may vary substantially and also have an impact on the values in the range (R) and fuel weight (W_f) columns. Similarly, the data in the drag-due-to-lift (k) column, the efficiency (e), and the values for W_m–the maximum weight and W_e–the empty weight represent some available average values and should be used with some caution.

D

Thrust Data for Performance Calculations

TURBOJETS

Engine data, for several sample engines, are given for J-60, J52, JT9D-3, JT8D-9, TF-30, TFE731-2, GE F404-400, FJ-44, Allison T-56 turboprop, and AVCO Lycoming I0-540 reciprocating engines. Although all these engines are somewhat dated and all have been superceded to give higher thrust and better fuel consumption, they do represent typical performance trends. As is clearly seen, the thrust and specific fuel consumption (SFC) curves vary widely with speed and altitude. Thus, there are no general-duty expressions available that would permit carrying out easy and simple performance calculations. However, several approximate expressions, shown below, have been used with some engineering success. As a first-order rough approximation, engine thrust can be scaled linearly (for similar engines) and the fuel consumption can be decreased by at least 5 percent.

The process consists of curve-fitting the thrust equation as a function of altitude, velocity, or both. Each engine may require its own special curve-fitting expression and may have to be accomplished in a piecewise fashion over the velocity and/or altitude range.

Jet Engines

For subsonic flight the simplest correlation is

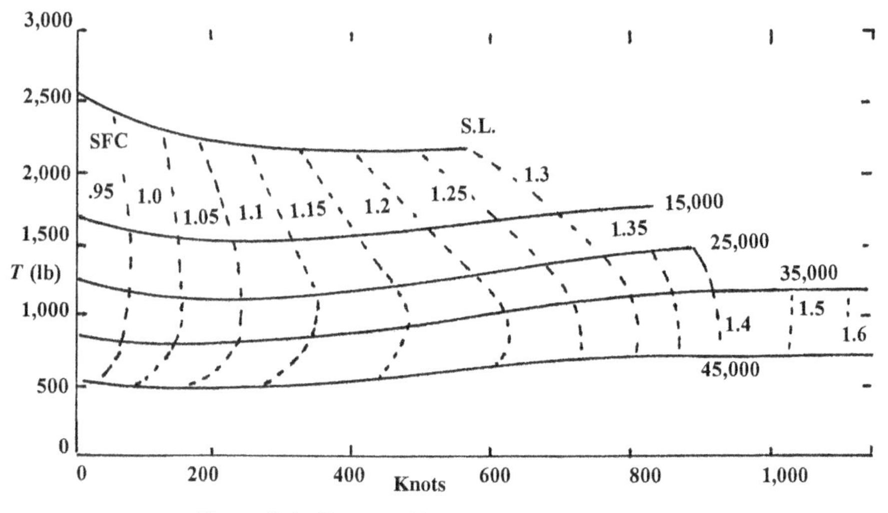

Figure D.1 Pratt and Whitney J-60 Turbojet Engine

$$T = T_{\text{ref}}\sigma \tag{D.1}$$

where T_{ref} may be taken as T_o, the sea level thrust value.

A somewhat improved expression is

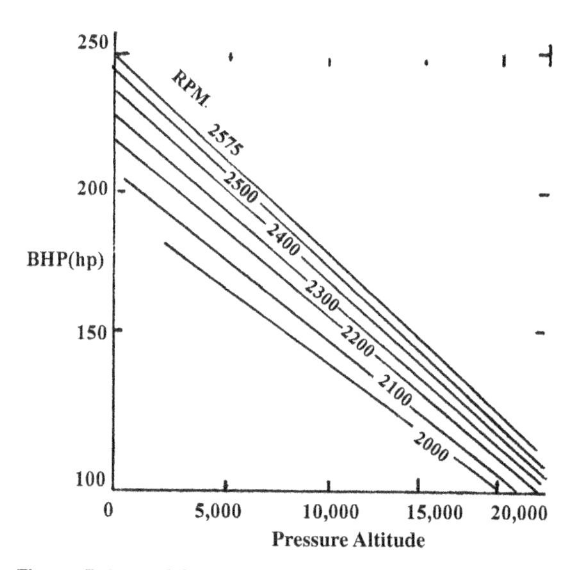

Figure D.2 AVCO Lycoming IO-540 Reciprocating Engine

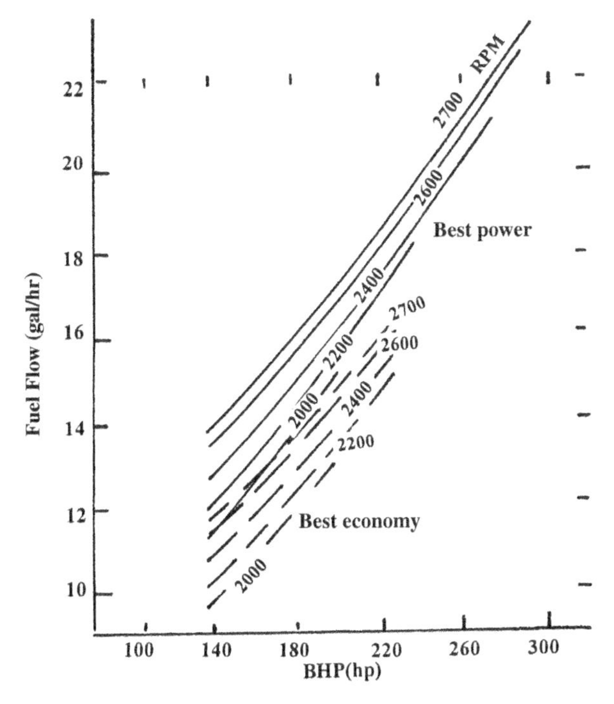

Figure D.3 AVCO Lycoming IO-540 Reciprocating Engine

$$T = T_{\text{ref}}\sigma^n \tag{D.2}$$

which is often written as

$$\frac{T}{T_o} = \sigma^n \qquad h < 36{,}089 \text{ ft}$$

$$= \sigma \qquad h > 36{,}089 \text{ ft}$$

A better correlation, but more cumbersome to curve fit and to use, is

$$\frac{T}{T_o} = (A + BV^2)\sigma \tag{D.3}$$

The advantage of Eq. D.3 lies in taking into account the realistic thrust variation with velocity. Usually this is not very significant at higher altitudes and velocities but may be 10 percent or more during

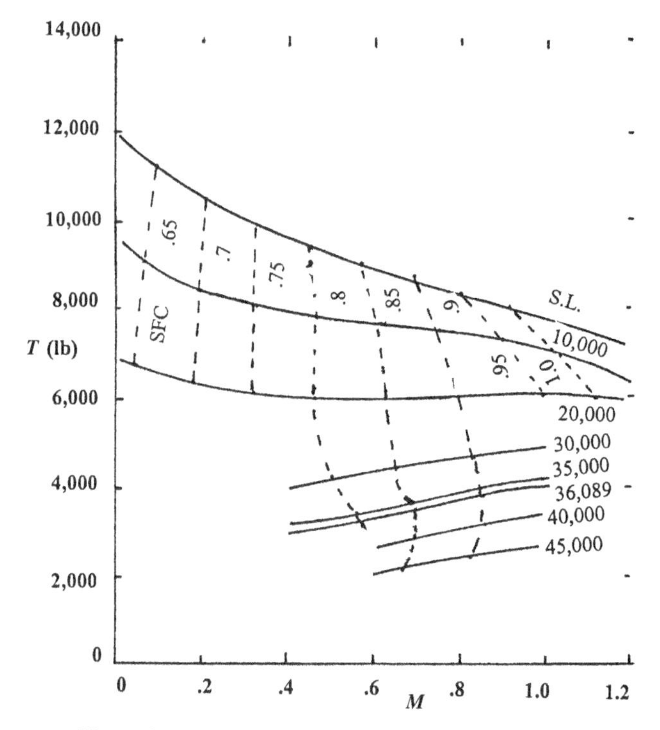

Figure D.4 Pratt and Whitney TF-30 Turbofan Engine

the take-off portion of flight (see TF-30 and JT9D data). At higher speeds, another correlation that has been used is

$$\frac{T}{T_o} = (1 + cM)\sigma \tag{D.4}$$

where, typically, $.25 < c < .5$.

Specific fuel consumption varies with both altitude and velocity and defies generalization with both of those parameters. It has been found that the velocity effect can be correlated for some engines, very approximately, by

$$\frac{TSFC}{TSFC_{\text{ref}}} = 1 + .5M \tag{D.5}$$

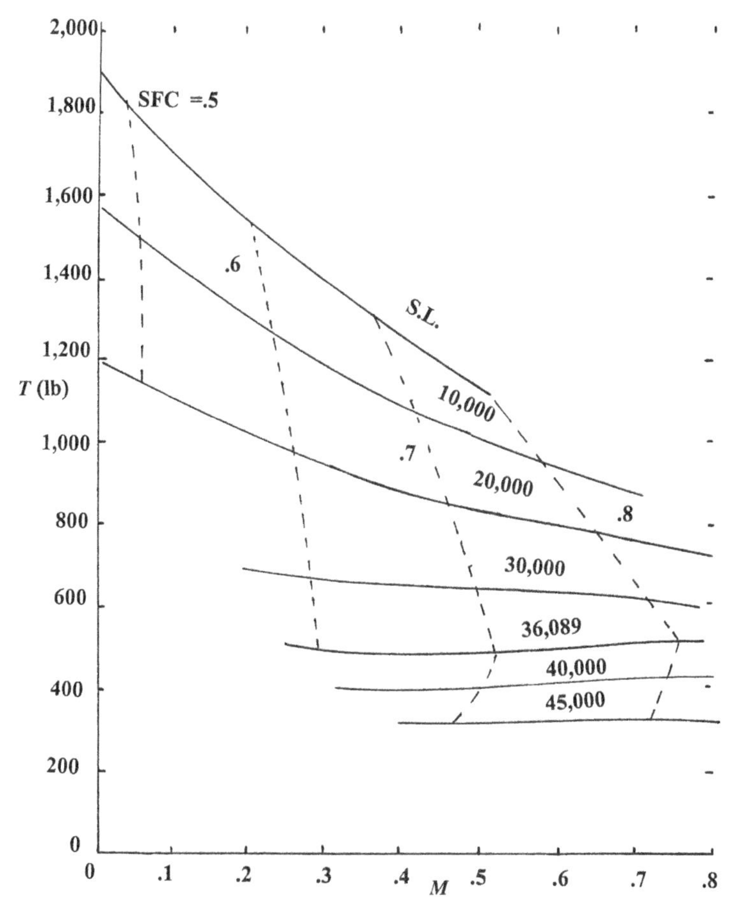

Figure D.5 Williams/Rolls FJ-44 Turbofan Engine

Reciprocating Engines

Reciprocating engines admit more generalizations:

- BHP is independent of velocity V.
- SFC tends to be independent of both velocity and altitude.

For engine brake horsepower, the commonly accepted altitude variation is

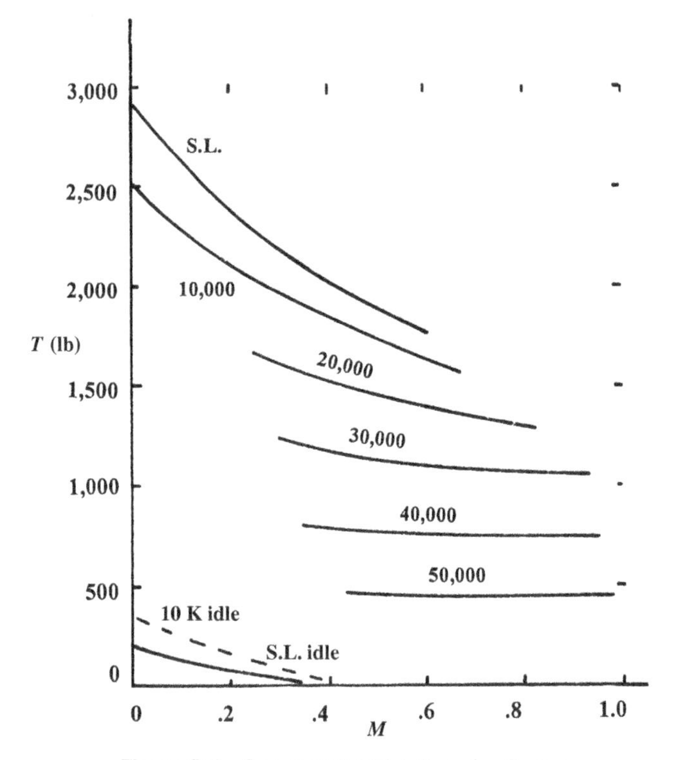

Figure D.6 Garrett TFE-731-2 Turbofan Engine

$$\frac{BHP}{BHP_o} = 1.132\sigma - .132 \tag{D.6}$$

where subscript o refers to the sea-level value.

For supercharged engines, it is assumed that BHP remains constant to at least 25,000 ft altitude. Correlations used for higher altitude supercharged engines are:

$$\begin{aligned} \frac{BHP}{BHP_o} &= \sigma^{.765}, \quad h < 36{,}089 \text{ ft} \\ &= 1.331\sigma, \; h > 36{,}089 \text{ ft} \end{aligned} \tag{D.7}$$

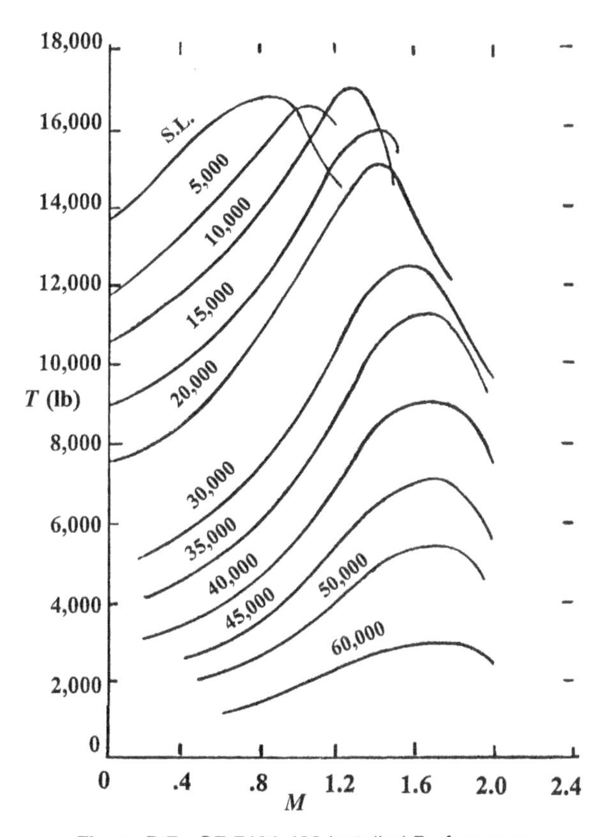

Figure D.7 GE F404-400 Installed Performance

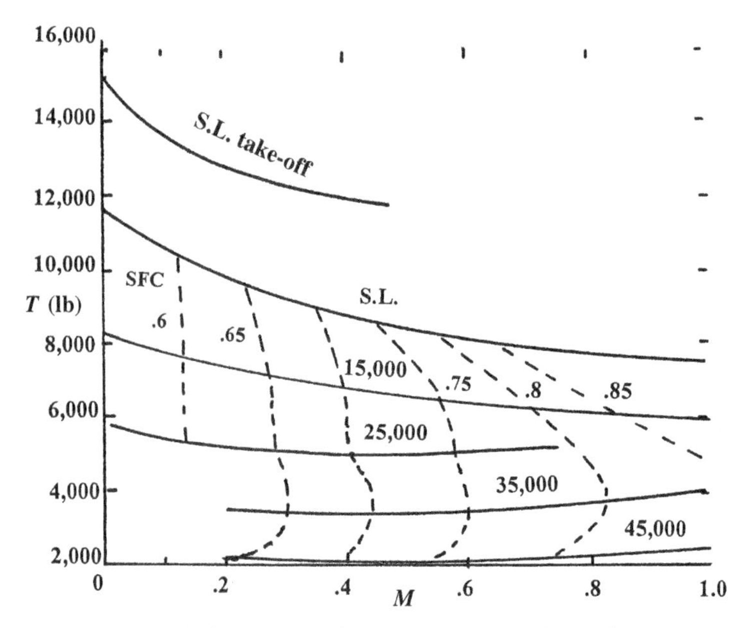

Figure D.8 Pratt and Whitney JT8D-9 Turbofan Engine

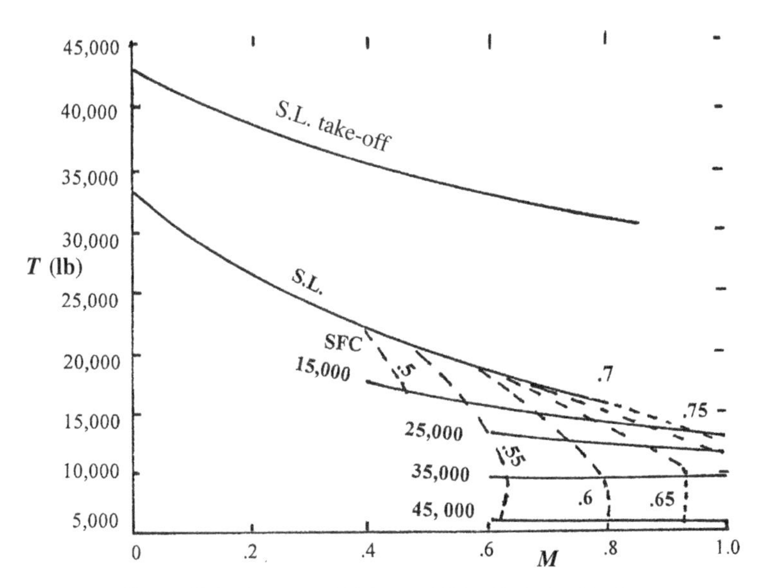

Figure D.9 Pratt and Whitney JT9D-3 Turbofan Engine

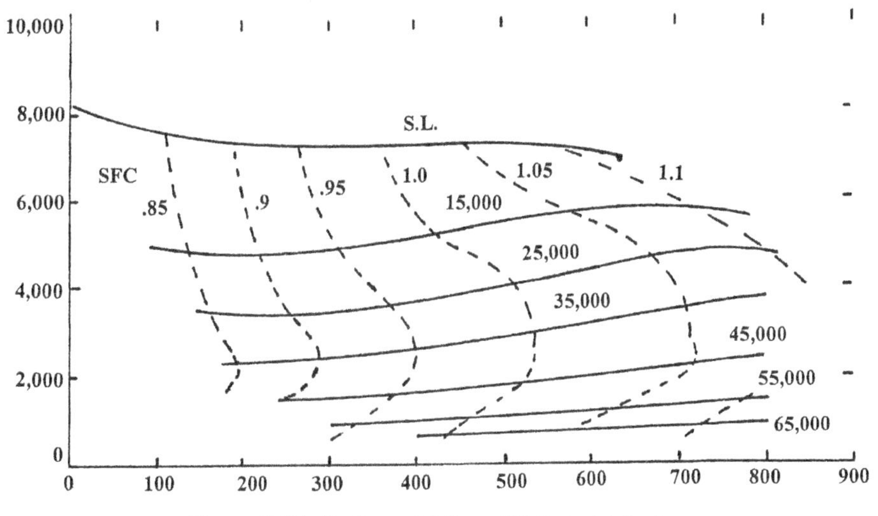

Figure D.10 Pratt and Whitney J52 Turbojet Engine

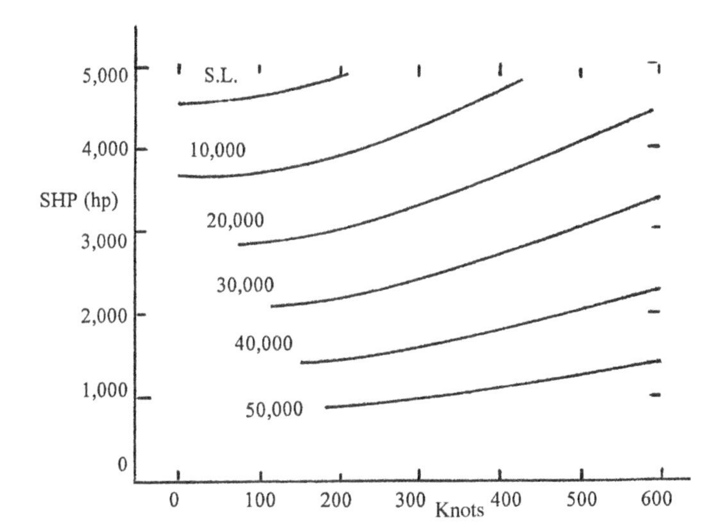

Figure D.11 Allison T-56-A Turboprop Engine, Horsepower

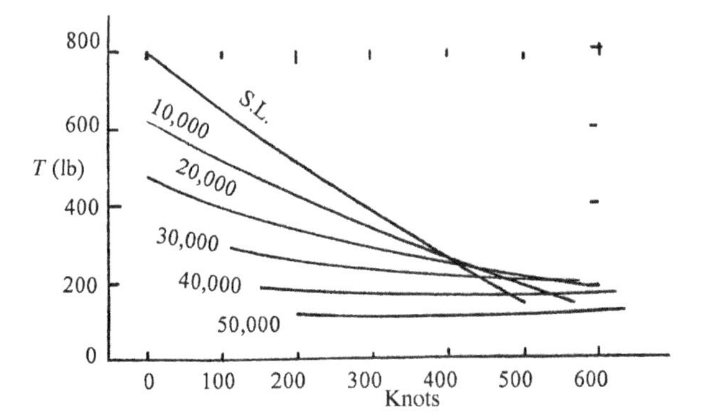

Figure D.12 Allison T-56-A Turboprop Engine, Thrust

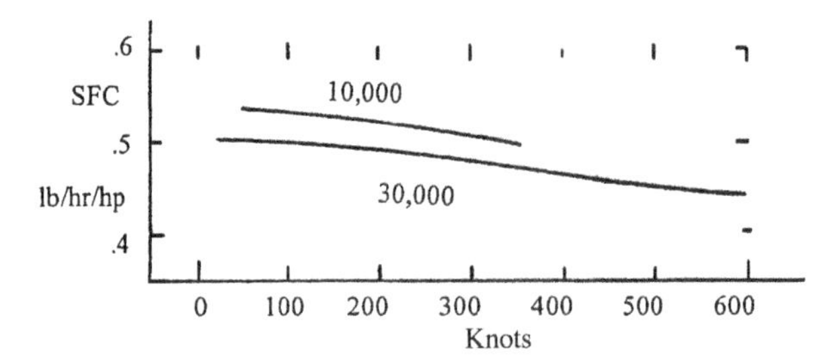

Figure D.13 Allison T-56-A Turboprop Engine, Specific Fuel Consumption

E

Some Useful Conversion Factors

Multiply	by	to obtain
meters	3.281	feet
$meters^2$	10.76	ft^2
Newton, N	.224	lb
	.102	kg_f
miles	5280	feet
	1.609	km
mph	1.467	ft/sec
	1.609	km/hr
	.869	knots
km/hr	.9113	ft/sec
	.2778	m/sec
	.6214	mph
	.54	knots
m/sec	3.281	ft/sec
	3.6	km/hr
	1.944	knots
	2.237	mph
knots	1.689	ft/sec
	1.852	km/hr
	.5144	m/sec
	1.151	mph

Bibliography

In Historical Order

1. Diehl, W. S. *Engineering Aerodynamics.* New York: Ronald Press, 1928.
2. Collection. *Handbook of Aeronautics.* London: H.M. Stationary Office, 1931.
3. Jones, B. *Elements of Practical Aerodynamics.* New York: Wiley, 1936.
4. Millikan, C. B. *Aerodynamics of the Airplane.* New York: Wiley, 1941.
5. Mises, R. *Theory of Flight.* Dover: McGraw-Hill, 1945.
6. Hemke, P. E. *Elementary Applied Aerodynamics.* New York: Prentice-Hall, 1946.
7. Wood, K. D. *Technical Aerodynamics.* Dover: McGraw-Hill, 1947.
8. Dwinnell, J. H. *Principles of Aerodynamics.* Dover: McGraw-Hill, 1949.
9. Perkins, C. D. *Airplane Performance, Stability and Control.* New York: Wiley, 1949. Hage, R. E.
10. Dommasch, D. O., Sherby, S. S., and Connolly, T. F. *Airplane Aerodynamics.* New York: Pitman, 1951–1967, 4 editions.
11. Miele, A. *Flight Mechanics.* Reading, MA: Addison-Wesley, 1960.
12. Houghton, E. L. *Aerodynamics for Engineering Students.* Arnold, 1960 and Brock, A. E.
13. McCormick, B. *Aerodynamics, Aeronautics, and Flight Mechanics.* New York: Wiley, 1979.
14. Lan, E., and Roskam, J. *Airplane Aerodynamics and Performance.* Roskam Aviation, 1980.

15. Hale, F. J. *Intro to Aircraft Performance, Selection, and Design.* New York: Wiley, 1984.
16. Asselin, M. *An Intro to Aircraft Performance.* AIAA, 1997.
17. Anderson, J. D. *Aircraft Performance and Design.* McGraw-Hill, 1999.
18. Eshelby, M. *Aircraft Performance, Theory and Practice.* AIAA, 2000.
19. Phillips, W. F. *Mechanics of Flight.* Hoboken: Wiley, 2004.

Two special references:

20. Rutowski, E. S. "Energy Approach to the General Aircraft Performance Problem." *Journal of the Aeronautical Sciences, 21* (March 1954).
21. Carson, B. H. *Fuel Efficiency of Small Aircraft.* AIAA Paper No. 80-1847, American Institute of Aeronautics and Astronautics, Washington, D.C., 1980.

Index